THE NEW GROVE
MODERN MASTERS

THE NEW GROVE
DICTIONARY OF MUSIC AND MUSICIANS

Editor: Stanley Sadie

The Composer Biography Series

BACH FAMILY

BEETHOVEN

HANDEL

HAYDN

MASTERS OF ITALIAN OPERA

MODERN MASTERS

MOZART

SCHUBERT

SECOND VIENNESE SCHOOL

WAGNER

THE NEW GROVE

Modern Masters

BARTÓK
STRAVINSKY
HINDEMITH

Vera Lampert
László Somfai
Eric Walter White
Jeremy Noble
Ian Kemp

MACMILLAN

First published in
The New Grove Dictionary of Music and Musicians,
edited by Stanley Sadie, 1980

First published in UK in paperback with additions 1984 by
PAPERMAC
a division of Macmillan Publishers Limited
London and Basingstoke

ISBN 0 333 37684 6

First published in UK in hardback with additions 1984 by
MACMILLAN LONDON LIMITED
4 Little Essex Street London WC2R 3LF
and Basingstoke

ISBN 0 333 37683 8

First American edition in book form with additions 1984 by
W.W. NORTON & COMPANY
New York and London

ISBN 0-393-30097-8 (paperback)
ISBN 0-393-01693-5 (hardback)

Printed in Hong Kong

Contents

List of illustrations

Illustration acknowledgments

We are grateful to the following for permission to reproduce illustrative material: Bartók Archives, Budapest (figs.1–9); M. I. Glinka State Central Museum of Musical Culture, Moscow (fig.11); Archives Théodore Strawinsky, Geneva (fig.13); André Meyer, Paris (fig.14); Boosey & Hawkes Ltd, London (figs.14, 20); Stravinsky Estate, New York (figs.14, 20); Victoria and Albert Museum, London (fig.16); Collection, The Museum of Modern Art, New York (gift of the artist) (fig.17); fig.18: photo Vera Stravinsky; Mrs Erich Auerbach, London (figs.19, 24; photo Erich Auerbach); Library of Congress, Music Division, Washington, DC (fig.20); B. Schott's Söhne, Mainz (figs.21, 23); Theatermuseum des Instituts für Theaterwissenschaft der Universität Köln (fig.22); Paul-Hindemith-Institut, Frankfurt (fig.23); CBS Records (cover).

General abbreviations

A	alto, contralto [voice]	inst	instrument, instrumental
a	alto [instrument]	ISCM	International Society
add, addn	addition		for Contemporary Music
appx	appendix		
ASCAP	American Society of	*Jb*	Jahrbuch [yearbook]
	Composers, Authors and	Jg.	Jahrgang [year of
	Publishers		publication/volume]
aut.	autumn		
		K	Köchel catalogue [Mozart;
B	bass [voice]		no. after / is from 6th edn.]
b	bass [instrument]	kbd	keyboard
b	born		
Bar	baritone [voice]	lib	libretto
bn	bassoon		
BWV	Bach-Werke-Verzeichnis	Mez	mezzo-soprano
	[Schmieder, catalogue of	movt	movement
	J. S. Bach's works]		
		ob	oboe
c	circa	orch	orchestra, orchestral
cimb	cimbalom	orchd	orchestrated (by)
cl	clarinet	org	organ
collab.	in collaboration with	ov.	overture
conc.	concerto		
cond.	conductor, conducted by	perc.	percussion
		perf.	performance,
d	died		performed by
db	double bass	PO	Philharmonic Orchestra
		pubd	published
edn.	edition		
eng hn	english horn	qnt	quintet
ens	ensemble	qt	quartet
facs.	facsimile	*R*	photographic reprint
fl	flute	rec	recorder
frag.	fragments	repr.	reprinted
		rev.	revision, revised (by/for)
gui	guitar	Rom.	Romanian
		Russ.	Russian
hn	horn		
Hung.	Hungarian	S	San, Santa, Santo [Saint];
			soprano [voice]
inc.	incomplete	sax	saxophone

SO	Symphony Orchestra	U.	University
Sp.	Spanish		
spr.	spring	v, vv	voice, voices
str	string(s)	va	viola
Swed.	Swedish	vc	cello
sym.	symphony, symphonic	vn	violin
T	tenor [voice]	WoO	Werke ohne Opuszahl
tpt	trumpet		[works without opus
Tr	treble [voice]		number]
trbn	trombone	ww	woodwind

Symbols for the library sources of works, printed in *italic*, correspond to those used in *Répertoire International des Sources Musicales*, Ser. A.

Bibliographical abbreviations

AcM	*Acta musicologica*
AMw	*Archiv für Musikwissenschaft*
BMw	*Beiträge zur Musikwissenschaft*
CMc	*Current Musicology*
FAM	*Fontes artis musicae*
GfMKB	*Gesellschaft für Musikforschung Kongressbericht*
Grove6	*The New Grove Dictionary of Music and Musicians*
IMSCR	*International Musicological Society Congress Report*
JAMS	*Journal of the American Musicological Society*
JMT	*Journal of Music Theory*
Mf	*Die Musikforschung*
ML	*Music and Letters*
MM	*Modern Music*
MMR	*The Monthly Musical Record*
MQ	*The Musical Quarterly*
MR	*The Music Review*
MT	*The Musical Times*
NOHM	*The New Oxford History of Music*, ed. E. Wellesz, J. A. Westrup and G. Abraham (London, 1954–)
NRMI	*Nuova rivista musicale italiana*
NZM	*Neue Zeitschrift für Musik*
ÖMz	*Österreichische Musikzeitschrift*
PNM	*Perspectives of New Music*
RdM	*Revue de musicologie*
ReM	*La revue musicale*
RIM	*Rivista italiana di musicologia*
SM	*Studia musicologica Academiae scientiarum hungaricae*
SMA	*Studies in Music*
SMz	*Schweizerische Musikzeitung/Revue musicale suisse*
SovM	*Sovetskaya muzïka*
ZMw	*Zeitschrift für Musikwissenschaft*

Preface

This volume is one of a series of biographical studies derived from *The New Grove Dictionary of Music and Musicians* (London, 1980). In its original form, the text was written in the mid-1970s and finalized at the end of that decade. For this print the text has been re-read and modified by the original authors and corrections and changes have been made. In the case of Bartók, the list of writings has been substantially supplemented in the light of recent publications, by László Somfai, who has also modified parts of the text. For Stravinsky, fuller details have been supplied in the list of works; further, an extended, critical bibliography, drawn up with the help of Richard Taruskin, replaces the previous one.

The fact that the texts of the books in this series originated as dictionary articles inevitably gives them a character somewhat different from that of books conceived as such. They are designed, first of all, to accommodate a very great deal of information in a manner that makes reference quick and easy. Their first concern is with fact rather than opinion, and this leads to a larger than usual proportion of the texts being devoted to biography than to critical discussion. The nature of a reference work gives it a particular obligation to convey received knowledge and to treat of composers' lives and works in an encyclopedic fashion, with proper acknowledgment of sources and due care to reflect different standpoints, rather than to embody imaginative or speculative writing about a composer's character or his music. It is hoped that the comprehensive work-lists and extended bibliographies, indicative of the origins of the books in a reference work, will be valuable to the reader who is eager for full and accurate reference information and who may not have ready access to *The New Grove Dictionary* or who may prefer to have it in this more compact form.

S.S.

BÉLA BARTÓK

Vera Lampert

László Somfai

CHAPTER ONE

Childhood and student years

Béla Bartók was born in Nagyszentmiklós in Hungary (now Sînnicolau Mare, Romania) on 25 March 1881. His father, also Béla Bartók (1855–88), was director of the agricultural school there and a keen amateur musician; he played the piano and the cello, composed short dance pieces, and founded a music society and an amateur orchestra in the town. The composer's mother, Paula Voit (1857–1939), who worked as a teacher, also played the piano. In such an environment Bartók's precocious musical gifts were quickly noticed. He had already shown talents for rhythm and memory when, on his fifth birthday, his mother gave him his first piano lesson. As a child he was quiet and withdrawn, often ill and plagued by a skin rash which began in infancy when he was vaccinated against smallpox; he later suffered from a bronchial condition.

The premature death of Bartók's father left the family in a precarious situation. His mother had to support the two children (Béla and his younger sister Elza, 1885–1955) by giving piano lessons, and in 1889 she took a teaching post in Nagyszőllős (now Vinogradov, USSR). It was there that, at the age of nine, Bartók produced his first compositions, most of them single-movement dances and some named after friends or members of the family; the 'Katinka' polka and the 'Irma' polka, both of 1891, were written for Katalin

1

Kovács and his aunt Irma. He also wrote a few programme pieces, notably *Radegundi visszhang* ('Echo of
Radegund', 1891), in memory of summer holidays with
his father, and *A Duna folyása* ('The course of the
Danube', 1890–94), which was inspired by a geography
lesson.

Keresztély Altdörfer, an organist from Sopron
who spent a few days in Nagyszőllős in 1890, predicted
a great future for the boy, and the next year he was
taken to Budapest for professional assessment at the
Academy of Music. There Károly Aggházy reassured
Bartók's mother about his talents and offered to take him
as a pupil, but she decided that he should not be separated from his family and should complete his inter-
mediate education in Nagyszőllős.

Having done that, Bartók entered the Gymnasium at
Nagyvárad (now Oradea, Romania) in 1891 and went
to live with his aunt, continuing his music studies with
the professional teacher Ferenc Kersch. According to
Bartók's mother, Kersch helped Bartók progress on the
piano but did not develop his understanding: 'He learnt a
very large number of pieces, but somehow superficially'.
In April 1892 his mother brought him back to
Nagyszőllős, breaking off his studies with Kersch, and
on 1 May that year he made his first public appearance
as both pianist and composer at a charity benefit for the
town; his programme included the first movement of
Beethoven's Waldstein Sonata and *The Course of the
Danube*. The concert was a great success, and Paula
Bartók determined to devote the next year (1892–3) to
her son's development. Taking leave of absence from her
post, she took the family to Pozsony (now Bratislava,
Czechoslovakia), a large town that could provide grea-

ter musical opportunities for the boy and where she hoped to find work. For that year Bartók's piano teacher was Ludwig Burger. Paula Bartók could find no permanent position, however, and so had to move to Beszterce (now Bistriţa, Romania) for 1893–4. It was such a small place that, according to her reminiscences, Bartók 'could not receive any musical training as he was the best pianist in town'. But a young violinist, Sándor Schönherr, was looking for an accompanist, and so Bartók was not entirely without the chance to exercise his talent. As Paula Bartók recalled, 'every week we had a concert at home; they played, among other things, Beethoven's violin sonatas'.

Finally, on 17 April 1894, the Bartóks settled in Pozsony again and Paula Bartók took a post at a teacher-training college. The young composer was at last to have five years during which his musical development was undisturbed by peregrinations. He had some excellent teachers – at first László Erkel, with whom he improved his piano technique, and later Anton Hyrtl, who gave him an excellent knowledge of harmony. At the same time he took an active part in the town's musical life; his friends included several amateur chamber musicians, and he attended concerts and the opera. In 1897 he played Liszt's Spanish Rhapsody in the town, in 1898 his Piano Sonata DD51, and (also in 1898, at a school concert) the last two movements of his Piano Quartet DD52. Through playing the organ at the Gymnasium chapel, he became acquainted with the repertory from Bach to Brahms, a widened experience reflected in his compositions of the period.

1898–9 was Bartók's last year in Pozsony. As he recalled in his autobiography of 1918, 'we were confronted with the question of which music school I should attend. In Pozsony, the Vienna Conservatory was then considered the sole bastion of serious musical training'. Accordingly, in December 1898 he travelled to Vienna and was seen by Hans Schmitt; he was auditioned, accepted and promised a scholarship. In the event, however, he took the advice of Dohnányi, who was four years his senior at the Pozsony Gymnasium and who, at that time, was his model of a composer–pianist. Dohnányi suggested he should attend the Budapest Academy of Music, and in January 1899 Bartók consulted Dohnányi's former teacher, István Thomán, who recognized the young man's talent and gave him a letter of recommendation to János Koessler, professor of composition at the academy. When he had completed his education in Pozsony he was auditioned for the academy and admitted to the second year for piano (under Thomán) and the second and third composition years (under Koessler); both were excellent teachers. In Thomán, one of the most gifted of Liszt's pupils, Bartók found not only a great teacher but a humane and supporting father figure. He provided the relatively impoverished student with scores, concert tickets, grants and recommendations (which helped to establish Bartók's career as a pianist), introducing him to celebrated artists and musicians.

Bartók's first year at the academy was successful, but in August 1900 he fell ill with pneumonia and, on doctor's advice, he stayed in Merano with his mother from November until early spring, returning to his studies in Budapest on 1 April 1901. He did not take the examinations that year and decided to repeat the third year of piano and the fourth of composition, completing

1. Béla Bartók aged 22

his studies in June 1903. Opinion at the academy was that he had an assured future as a virtuoso pianist; composition was expected to be a secondary pursuit. He was an excellent sight-reader and chamber player, and was often sought by his teachers as an accompanist. On 21 October 1901 he played Liszt's B minor Sonata at a student concert, his début in Budapest; later in the year he was invited to accompany the violinist Jenő Hubay. The following year he attracted attention by playing *Ein Heldenleben* at the academy and in January 1903 in Vienna at the invitation of the Tonkünstlerverein. In his last year as a student (1903) he performed at the academy on ten occasions and gave his first solo recital (13 April) in his home town of Nagyszentmiklós. He was gaining a reputation in Hungarian musical circles and soon the salons of the great patrons were to open to him. At the home of Emma Gruber (née Sándor, later Kodály's wife), for instance, many of his pieces had their first hearing and received their first criticisms; and Bartók dedicated to her the second of the Four Piano Pieces (1903) and the Rhapsody op.1 (1904).

During his Budapest years Bartók broadened his musical interests. He got to know the works of Wagner: previously he had seen only *Tannhäuser* in Pozsony; now he studied the *Ring*, *Tristan* and *Die Meistersinger* as well as Liszt's scores. But none of his studies suggested a new compositional direction. Koessler, who was a devoted follower of German Romantic ideals, helped Bartók to a mastery of the requisite techniques, but this did not stimulate his creativity. All he composed during the first year and a half, apart from a few exercises, were pieces inspired by his friends. One, the fourth of the *Liebeslieder* (1900), employs a theme by Felicie Fábián, his friend at the academy. The two had become acquainted at the beginning of their studies and had developed a close relationship, working together, sharing history notes and criticizing each other's compositions. When Bartók fell ill it was Fábián who introduced his compositions at seminars. In 1900 Bartók composed a scherzo for piano (DD63) based on a motif made up of their initials, F–F–B–B, and in January 1901 he completed a set of piano variations on a theme by Fábián (DD64); the scherzo is dedicated to her. At the beginning of 1903 their friendship was interrupted when Fábián departed for Vienna to continue her studies. Two small violin pieces, the Duo DD69 and the Andante DD70, both written in November 1902, were connected with another friend, Adila d'Arányi (later Adila Fachiri).

CHAPTER TWO

Discoveries: Strauss, Folk music, Debussy

Bartók's discovery of Strauss marked a decisive change in his career as a composer. He was so struck by *Also sprach Zarathustra*, the Budapest première of which he attended in February 1902, that he enthusiastically began a study of Strauss's scores, memorizing *Ein Heldenleben*; he later wrote an article introducing the *Sinfonia domestica* for *Zeneközlöny* (February 1905). His enthusiasm for composition returned under Strauss's influence, already noticeable in the partly orchestrated Symphony DD68 of late 1902. The other great influence on his music at this time was the increasing nationalist feeling of Hungarian independence; he took to wearing national dress and opposed the everyday use of German by his family. His compositions of this period have a distinctly Hungarian tone: the Four Pósa Songs DD67 (1902), the Four Piano Pieces DD71 (1903), the Violin Sonata (1903), the symphonic poem *Kossuth* DD75 (1903) and the Piano Quintet DD77 (1903–4).

Summer 1903 Bartók spent in Gmunden, where he had some piano lessons with Dohnányi. In the autumn he went for a while to Berlin to study and give concerts, leaving from time to time to give concerts in Vienna, Budapest and Pozsony; some of his own pieces were included in his programmes. He was introduced as a composer too: Rudolf Fitzner gave the first performance of the Violin Sonata in Vienna in 1904, and *Kossuth* was performed that year under Hans Richter in Budapest and Manchester. Also in 1904 the Pósa Songs and the Four Piano Pieces were published by Bárd. For six months from April Bartók retired to the countryside (at Gerlice, Gömör, Slovakia), preparing for concerts, and composing; his new repertory included several works of Liszt. On 21 November 1904 the Piano Quintet was given its first performance by the Prill Quartet in Vienna. Finally, in 1904 Bartók completed the works to which he later gave the ascriptions opp.1 and 2, the piano Rhapsody (the version 'for orchestra and piano' remained in his repertory until 1936) and the Scherzo for piano and orchestra.

Bartók spent the first few months of 1905 in Vienna, where he composed the First Orchestral Suite and gave a piano recital (February); in March at Budapest he played Liszt's *Totentanz* with great success. The summer found him in the country again, at Vésztő in the Békés county, preparing for the quinquennial Rubinstein competition in Paris. Since he was entering as both composer and pianist, he submitted the Rhapsody and the recently completed Piano Quintet for consideration. The latter was not played because the musicians had insufficient rehearsal time, but after a long search Bartók found a violinist to play his Sonata. Lack of money prevented the composition prize being awarded, and only the first prize for piano was given – to Backhaus. Of the five composer entrants only Bartók and Brugnoli received certificates. The competition did not fulfil his expectations, but Bartók's visit to Paris was a significant experience in other ways. In November 1905 he played *Totentanz* in Manchester and the First

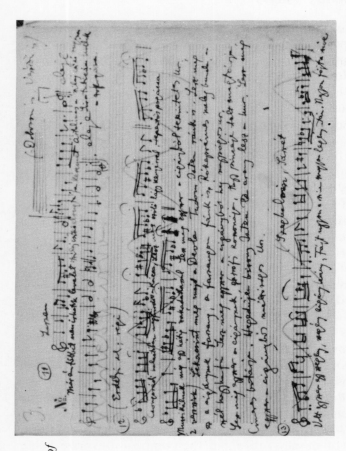

2. Page from Bartók's field notebook: draft of Hungarian folk melodies

10

Suite was introduced by the Vienna PO (29 November). The following March and April he accompanied the child prodigy violinist Ferenc Vecsey (just 13 years old) on a tour of Spain and Portugal. But in 1906 he produced no new composition, even putting aside the Second Orchestral Suite, begun in 1905.

Until this period Hungarian nationalist composers had used folklike popular songs as their models. Bartók soon recognized that they were not autochthonous Hungarian folksongs (autobiography, 1918). In 1904 he made his first notation of a Hungarian peasant song, sung by a young girl in the Gömör district, and this discovery drew his attention to the treasury of indigenous song which might contain innumerable ideas for 'serious' composition. In 1905 he made more notations in Békés, and he made contact with Kodály, whom he had previously known superficially and who had recently published his first study of folk music. He found in Kodály an expert on the subject and a helpful colleague. Theirs was to be a lifelong collaboration, the first product of which was the publication of *Magyar népdalok* ('Hungarian folksongs', 1906), containing ten settings for voice and piano by each composer. Their primary intention was to popularize folksongs, but it took 32 years to sell 1500 copies according to the second edition (1938). The poor reception of the first volume prevented the publication of a second, but Bartók had already harmonized ten more folksongs. So they began to devise a scientific research method and analytical system. From 1906, using an Edison phonograph, Bartók travelled annually all over Hungary doing fieldwork and making recordings.

In January 1907 Bartók was appointed to a piano

professorship at the Budapest Academy of Music, a position he welcomed since it enabled him to settle in Hungary, and continue his folklore research. At the same time his financial situation became more secure, though on a modest scale. He soon extended his research to other nationalities; from 1906 he was collecting Slovak songs, in 1908 he notated his first Romanian folksong. In Transylvania in 1907 he had discovered the pentatonic origins of Hungarian peasant song. The discoveries of the first collecting tours inspired the first folksong arrangements, *Három Csík megyei népdal* ('Three Hungarian folksongs from the Csík district', 1907), the Four Slovakian Folksongs (*c*1907), some of the Eight Hungarian Folksongs, and the piano pieces *Gyermekeknek* ('For children', 1908–9). The fifth of the *Vázlatok* ('Sketches', 1908–10) is a Romanian folksong arrangement, and other works of the period with some relation to folk music include the 14 Bagatelles (1908), the Ten Easy Pieces (1908), the *Két elégia* ('Two elegies', 1908–9), the *Három burleszk* ('Three burlesques', 1908–11) and the *Allegro barbaro* (1911).

Under the influence of Kodály, who had spent a few months in Paris in 1907, Bartók became interested in Debussy's music (he already had a few scores), in which he found elements similar to those in the folk music with which he had been working. For a short time he also took an interest in Reger. The first of Bartók's works to synthesize the influences of peasant song and art music were the First String Quartet (1908) and the orchestral *Két kép* ('Two pictures', 1910). The original version of a two-movement violin concerto (Sz 36; later called no.1, contrary to Bartók's wishes) was not published in his lifetime, and he even abandoned the idea of having it

3. Bartók and Kodály in 1908

performed, though originally he had offered it to Henri Marteau. It had been written for Stefi Geyer, a talented young violinist to whom Bartók wrote uncharacteristically long and revealing letters, discussing matters of religion, philosophy and art; 'If I would cross myself I would say "in the name of Nature, Art and Science" ', he wrote to her on 6 September 1907. In another letter he called the concerto's theme the 'Stefi Geyer leitmotif'; he dedicated the work to her and gave her the score. The theme recurs in several piano works of the period: the 'Dedication' from the Ten Easy Pieces is built on the Stefi Geyer motif. Bartók wrote the 13th of the 14 Bagatelles, a *Lento funebre* sub-titled 'Elle est morte', on the day he received the parting letter from Geyer (14 February 1908); the first elegy was composed in the same month. The first version of the 14th bagatelle, the presto waltz 'Ma mie qui danse', was made on 20 March 1908; an orchestrated version of this piece functions as the 'grotesque' movement of the *Két portré* ('Two portraits'). Busoni greatly admired the experimental style of the 14 Bagatelles, and shortly afterwards invited Bartók to conduct the orchestra in Berlin.

In autumn 1909 Bartók married his pupil Márta Ziegler. She had inspired the first of the *Sketches*, 'Portrait of a Girl' (1908), and the 'Quarrel' from the *Three Burlesques* (also 1908); later she was to be the dedicatee of the opera *A Kékszakállú herceg vára* ('Duke Bluebeard's Castle'). Their son Béla was born in 1910.

CHAPTER THREE

Neglect and success, 1910–20

Bartók's first 'composer's evening' was on 19 March 1910 and included the première of the First Quartet (by the Waldbauer) as well as performances of the Piano Quintet and various piano pieces. A week earlier, at a concert in Paris introducing young Hungarian composers, Bartók had played a selection from the Bagatelles. In April 1911, at the instigation of Bartók, Kodály and a few others, the new Hungarian Music Society was formed with the intention of performing contemporary Hungarian works; Bartók himself played in their first concert, when the programme included works by Debussy, Ravel and the two leading sponsors. They planned a new orchestra but failed because of lack of money and interest.

Bartók moved away from the capital in 1912 to the suburb of Rákoskeresztur, where he rented a flat. He had withdrawn from all public musical activity after a number of fiascos: *Bluebeard's Castle* had been rejected by the jury of the national opera competition, and his publishers Rózsavölgyi and Rozsnyai showed no interest in new works, though they had issued a dozen of his pieces between 1908 and 1912. (They did, however, ask him to prepare teaching editions of Haydn, Mozart, Beethoven and Bach, which became the standard pedagogical texts in Hungary.) For the next few years his output was once more to be small. The Four Pieces

4. Bartók with Kodály (front right) and the Waldbauer Quartet (from the left: Jenő Kerpely, Imre Waldbauer, Antal Molnár, János Temesváry)

16

for orchestra, composed in 1912, were left for nine years before being orchestrated; the only works dating from 1913 are some easy piano pieces for Reschofsky's piano school.

'With more enthusiasm than ever, I devoted myself to studies in musical folklore. I planned, considering our modest circumstances, some brave journeys' (autobiography, 1921). Bartók could not realize his plans to visit Russia in 1912, but in 1913 he visited Biskra in north Africa. Besides a few Ruthenian, Bulgarian and Serbian items, by 1918 he had collected 2721 Hungarian, about 3500 Romanian and 3000 Slovak folksongs. He also began to systematize his collections for publication, using a refined version of Ilmari Krohn's classification system.

In 1911 Bartók was in touch with Kiriac, who used his influence to have Bartók's first ethnomusicological monograph published in Bucharest (1913). In the same year Bartók and Kodály suggested that a collected edition of Hungarian peasant songs be prepared (its publication, as *Corpus musicae popularis hungaricae*, was delayed until after World War II). Bartók also lectured in Budapest on the musical dialects of the Hunyad district Romanians, with peasant singers present to demonstrate; his book on the folk music of Máramaros was not published until 1923 because of the war. Also, the war years left him free to compose, since his delicate physical condition made him unfit for service. In 1915 he wrote a number of pieces based on Romanian folksong (the Piano Sonatina and various arrangements for piano, female chorus, and voice and piano); and in 1917 he wrote the choral *Tót népdalok* ('Slovak folksongs'), for a concert at the Musikhistorische Zentrale, Vienna,

17

in January 1918. He also completed the *Nyolc magyar népdal* ('Eight Hungarian folksongs') for voice and piano and the *Tizenöt magyar parasztdal* ('15 Hungarian peasant songs') for piano. Other works of this period, the Piano Suite (1916) and the Second Quartet (1915–17), show the influence of Arab music. Two song cycles also appeared in 1916. The first, op.15, was not published during Bartók's lifetime and uses four poems of Klára Gombossy (though she is not credited on the score). Bartók met her in September 1915 when she was 15, and she accompanied him on collecting expeditions; their friendship lasted for a year. The other cycle is to poems of Ady.

Another work to benefit from this renewed activity was the ballet *A fából faragott királyfi* ('The wooden prince'), which was completed in 1917; its première was on 12 May that year at the Budapest Opera House under Egisto Tango, to whom Bartók dedicated the score. The successful production of the ballet satisfied not only Bartók but also the audience; it changed the public attitude towards his music. *Bluebeard's Castle* was belatedly introduced in the following year, again under Tango, and on 3 March 1918 the Second Quartet was played for the first time by the Waldbauer.

Under the Republic of Councils (which lasted only four months from 21 March 1919) Bartók was brought back into public life. With Kodály and Dohnányi he joined a music committee directed by Béla Reinitz, to whom in 1920, when the latter was threatened with imprisonment, he dedicated the Five Ady Songs op.16 'with true friendship and love'. The reforming aims of the committee were primarily concerned with education, and they also planned to institute a folk

5. *Autograph sketch of the beginning of the third movement of Bartók's String Quartet no.2*

19

music department within the National Museum, with Bartók as director. A second 'composer's concert' was arranged for Bartók, the programme including the premières of the Ady cycle, the Piano Suite and the Studies (1918) for piano. It was at this time that Bartók completed the piano score of his second ballet, *A csodálatos mandarin* ('The miraculous mandarin').

After the war, the political situation in Hungary prevented Bartók from travelling where he wanted for folksong research; the Treaty of Trianon (June 1920) annexed the northern part of the country to Czechoslovakia and the southern (Transylvania) to Romania. In spring 1920 Bartók was accused of lack of patriotism, the case arising from the delayed German publication of his study *Der Musikdialekt der Rumänen von Hunyad*. He contemplated emigration. February and March he spent in Berlin, participating in two concerts for the Neue Musikgesellschaft and inquiring into the possibilities of settling; he also considered Vienna, Transylvania and London. But he found himself unable to leave Hungary, and before the end of 1920 he was back in Budapest, where he stayed for two years in the villa of József Lukács, father of György Lukács.

CHAPTER FOUR

Concert tours

By this time Bartók was established as an international figure. Universal Edition had begun to publish his works in 1918; Schoenberg's Verein für Musikalische Privataufführungen had presented several of his works (the 14 Bagatelles, the First Quartet, the Rhapsody op.1 etc); the *Musikblätter des Anbruch* published a special Bartók issue for his 40th birthday; the *Revue musicale* marked the occasion with an article by Kodály; and in London his works were studied and promoted by Calvocoressi, Gray and Heseltine. Bartók himself entered into the musical life of Europe in the 1920s. He visited London and Paris in 1922, playing his First Violin Sonata with Jelly d'Arányi and meeting Stravinsky, Szymanowski, Ravel, Milhaud, Poulenc and Satie. The next year he was back in London with d'Arányi and his Second Sonata. From then on he made regular recital tours (generally two or three each season) of Germany, the Netherlands, England, Switzerland, Italy, Czechoslovakia and Romania. Apart from giving solo recitals, he accompanied the celebrated Hungarian violinists of the day: Waldbauer, Szigeti, Székely, Gertler and others. In 1928 he made the first recordings of his music which he played with Maria Basilides, Ferenc Székelyhidy and Vilma Medgyaszay.

Bartók's European success assisted his standing in Hungary. In 1922 the Improvisations on Hungarian

Peasant Songs for piano (1920) and the Four Pieces for orchestra (1912–21) were given in Budapest, and Bartók was commissioned to write a work in celebration of the 50th anniversary of the union of Buda and Pest. The resulting piece, the *Tánc-Suite* ('Dance suite') together with commissioned works from Kodály and Dohnányi, was played at a concert in Budapest in November 1923. That year also brought changes in Bartók's personal life: he divorced his first wife in the autumn and married his pupil Ditta Pásztory, who was to be the dedicatee of *Falun* ('Village scenes'), 'Az éjszaka zenéje' ('Night music') from the piano suite *Szabadban* ('Out of doors'), the Piano Sonata and other works. His second son, Péter, was born in 1924.

During the 1920s Bartók continued work on his folksong collection and prepared some for publication. In 1921 he collaborated with Kodály on a volume of 150 Transylvanian songs; his study of Hungarian peasant song was published in 1924. A collected edition of 3000 Slovak folksongs was prepared between 1922 and 1928; and in 1926 he wrote an extensive study of Romanian *colinde* (Christmas songs). He also published a number of short articles on folksong, on the influence of folksong on art music, and on the problems of 20th-century music; these appeared both in Hungarian and leading foreign journals. Other writings included reports on Budapest musical life for the *Musical Courier* and *Il pianoforte*, and dictionary entries.

After *Village Scenes* Bartók composed nothing for 18 months. Concert tours and teaching took up most of his time during this period, leaving only the summers for composition. Summer 1925 was taken up editing *colinde*. It depressed him that *The Wooden Prince* and

Bluebeard's Castle could not be performed at the Budapest Opera because Béla Balázs, the author of the texts, was politically compromised and in voluntary exile (the works were not to be given again until 1935 and 1936 respectively). The première of *The Miraculous Mandarin* was repeatedly delayed but finally took place at Cologne on 27 November 1926, when the audience left in a state of outrage; further performances were banned. However, the Prague production of the next year was well received. The Dance Suite, too, proved a success, with performances under Talich in Prague and Budapest; in the same season in Germany alone it was played over 50 times.

Bartók's increasing international renown as a performer encouraged him to compose new piano works, and in 1926 alone he produced the First Piano Concerto, the Sonata, *Out of Doors* and the Nine Little Pieces; the concerto was introduced at the 1927 ISCM Festival, and on 16 July that year Bartók played his Sonata at a concert in Baden-Baden which also included Berg's Lyric Suite. He spent summer 1927 working on his Third Quartet, which had its first performance, again given by the Waldbauer, in London on 19 February 1929. The Waldbauer also introduced the Fourth Quartet, in Budapest on 20 March that year. The Second Piano Concerto followed in 1931 and was played by Bartók more than 20 times between 1934 and 1941. The demand for virtuoso pieces and easily understood music increased, and he made orchestral arrangements, also writing the violin rhapsodies for Szigeti and Székely. Besides all this, he composed the 44 Violin Duos on commission from Doflein as teaching pieces, and he followed Kodály's *Magyar népzene* ('Hungarian folk music') series in writing a multi-volume set of folksongs for voice and piano, *Húsz magyar népdal* ('20 Hungarian folksongs', 1929). The *Magyar népdalok* (five 'Hungarian folksongs', 1930) for mixed voices and the *Székely dalok* ('Transylvanian songs', 1932) for six-part male chorus, written for the Bratislava choir named after him, were his last large-scale transcriptions and his finest.

The *Cantata profana*, Bartók's most important non-operatic vocal work, dates from the same period (1930). Its text, taken from Romanian *colinde*, was translated and arranged by him (see his *Melodien der rumänischen Colinde* (*Weihnachtslieder*), 1935, text no.4*a–b* in Dille, 1968), and he planned to compose two or three more cantatas, noting that 'a common idea should unify these works, but each could be performed separately'. Only a fragment of a continuation appears to survive, with a text beginning: 'Once upon a time three different worlds were in rivalry; three different worlds, three different countries'. This suggests that the theme was to have been the brotherhood of nations, a belief dear to Bartók. In a letter of 1931 to Octavian Beu he wrote: 'My true conviction . . . is of the brotherhood of nations'. The *Cantata profana* had its première in London on 25 May 1934; Dohnányi conducted it in Budapest on 10 November 1936.

Between December 1927 and February 1928 Bartók made a tour of the USA. He had intended to make his New York début with the First Piano Concerto under Mengelberg, but lack of rehearsal time forced him to substitute the Rhapsody op.1. He finally gave the concerto in New York at the end of his tour (13 February), in a concert conducted by his pupil Fritz Reiner. His recitals included Kodály's works as well as his own; he generally introduced them with a lecture on new Hungarian music. The American tour was followed by a visit to the USSR (December 1928 to January 1929) and a concert in Basle (January 1929), where he met Paul Sacher. He also took part in ISCM events, lectured at the 1928 congress on folk music in Prague and attended an Arab music conference in Cairo in 1932.

On his 50th birthday Bartók was honoured with the medal of the Légion d'honneur and the Corvin Medal, but he did not attend the ceremony of the latter. Revivals of *The Wooden Prince* and *Bluebeard's Castle*, and the Budapest première of *The Miraculous Mandarin* had been planned for the occasion, but did not materialize. For the next six years Bartók declined to play his works in Budapest and rarely took part in concerts, though he made exceptions for performances with old friends (Waldbauer, Szigeti, Dohnányi and others), a few performances of *Totentanz* and recitals for broadcasting. He did not break his interdict, however, even for the Budapest première of the Second Piano Concerto (2 June 1933), given by Louis Kentner.

All this time Bartók continued to serve at the Budapest Academy of Music. He was a reluctant teacher and would have preferred a research post; but he taught a number of excellent musicians, including Kósa, Arma and Reiner. In autumn 1934 he was finally relieved of his teaching duties at the academy; the Hungarian Academy of Sciences commissioned him to prepare the publication of the Hungarian folksong collection planned since 1913, and he systematized its some 13,000 items. Besides Hungarian folksongs he revised and arranged a collection of 2500 Romanian folksongs, and he prepared further foreign collections for future analysis; what he had learnt of Bulgarian rhythms had made him revise his approach (for a survey of his research on the interrelationship between Hungarian, Slovak, Ruthenian and Serbo-Croatian folk music, see his essay *Népzenénk és a szomszéd népek népzenéje*, 1934). In November 1936 he was in Turkey, and with Adnan Saygun he collected folksongs in Anatolia. In a sense, however, his teaching work continued in lectures on folk music at home and abroad, and in the composition of *Mikrokosmos* (1926–39), a set of six volumes of progressive studies begun when his second son Péter started to tackle the piano. Furthermore, the first four volumes of two- and three-part choruses (1935) were intended for school use, stimulated by the example of Kodály. In May 1935 Bartók was appointed a member of the Hungarian Academy of Sciences.

In the 1930s Bartók's important compositions were written to commission. The Fifth Quartet (1934) was composed for Elizabeth Sprague Coolidge, the Music for Strings, Percussion and Celesta (1937) and the Divertimento (1939) for Paul Sacher, the Sonata for two pianos and percussion (1937) for the Basle ISCM group, the Violin Concerto (1937–8) for Székely, and *Contrasts* (1938), commissioned by Goodman, for Szigeti and Goodman. Bartók and his wife gave the first performance of the Sonata in Basle on 16 January 1938, her début as a pianist.

CHAPTER FIVE

Last years

The threat of fascism had concerned Bartók from the first, and he felt obliged to protest against it. In 1931, when Toscanini was under attack from the fascists, he wrote on behalf of the Hungarian section of the ISCM 'to protect the integrity and autonomy of art'. After the first performance of the Second Piano Concerto under Rosbaud in Frankfurt (23 January 1933) he never again played in Germany, and in 1937 he forbade broadcasts of his music in Germany and Italy. He renounced his membership in the Austrian Performing Rights Society as it was Nazi-inclined and joined the London branch; he left Universal Edition, and from 1937 his works were published by Boosey & Hawkes.

Bartók was being attacked in the Hungarian and Romanian newspapers, and he began to look to his Swiss friends for help: 'The political situation in Hungary becomes more and more crooked ... at least my manuscripts should be somewhere safe' (to Müller-Widmann, 13 April 1938). The next year he directed that his papers be sent to London, but would have preferred them to go even further, to the USA. Meanwhile he considered emigration. His unfinished folk music research and his aged mother tied him to Budapest, and it was only after her death (December 1939) that he began to inquire about the possibilities of settling elsewhere. He made a tour of the USA (11 April to 18 May 1940), during which he gave a (recorded) concert with Szigeti in the Library of Congress, recorded *Contrasts* with Szigeti and Goodman, and lectured on folk music; a longer visit was scheduled for the autumn. On 8 October 1940 he gave his farewell concert in Budapest, playing Bach's A major Concerto, a selection from *Mikrokosmos* and Mozart's Two-Piano Concerto (with his wife, who made her solo début in κ413); the conductor was János Ferencsik. A few days later the Bartóks left Hungary, travelling through Switzerland to Lisbon, where they embarked for the USA on 20 October. Bartók's last significant European composition had been the Sixth Quartet (1939), which had its first performance by the Kolisch Quartet in New York (20 January 1941).

On arriving in New York the Bartóks moved into a hotel and then rented a flat for a few months in Forest Hills, Long Island. From May 1941 for two years they lived in Riverdale, in the Bronx, whence they moved to Bartók's last home, an apartment on West 57th Street. The couple gave duo concerts from their first arrival, but they were not a resounding success and their appearances did not provide a sufficient income.

In spring 1940 Bartók learnt of the existence of the Parry collection of some 2600 discs of Yugoslav folk music, as yet unclassified, at Harvard University. He showed great interest in this material, since it could be presumed to relate to his own work, and the Ditson Foundation of Columbia University made it possible for him to work on the collection as visiting assistant in music. Columbia awarded him an honorary doctorate (November 1940) and he was employed from March 1941 to the end of 1942 at a salary of $3000 a year, though since the post was renewable each term he could never feel secure. He did, however, gain pleasure from working on the Parry collection, and his findings were published. At the same time he maintained contact with the University of Washington at Seattle, where he had had the offer of lecturing for a year. Because of the distance involved, however, and the problems of moving house, he had delayed his decision; moreover, he was not interested in working on American Indian folk music.

During his first two years in the USA Bartók completed no new work. He did arrange the Sonata for two pianos and percussion as a double piano concerto (December 1940) and he also transcribed the Second Suite for two pianos, both for concerts with his wife. Most of his time was devoted to ethnomusicological work: while studying the Parry collection he was editing his own copious Romanian material, which by 1942 he had prepared for publication. However, despite a previous agreement, the New York Public Library found the prospective printing costs of the two-volume work too great and rejected it. Bartók then prepared a smaller study of his Turkish material, collected in 1936, but this too was rejected, and he deposited the manuscripts with Columbia University.

From April 1942 Bartók's health began to decline: he suffered from high temperatures, and a series of medical examinations produced no firm diagnosis. He gave his last public performance on 21 January 1943, playing with his wife in the USA première of the Concerto for Two Pianos at a New York PO concert under Reiner. The next month he gave three lectures at Harvard, but he was completely exhausted by them and had to undergo a further course of medical tests. This time his illness was diagnosed as polycythaemia. His former pupil Ernő Balogh asked the American Society of Composers, Authors and Publishers (ASCAP) to finance treatment, to which they agreed. In May Koussevitzky commissioned an orchestral piece; Bartók never learnt that the suggestion originated with Szigeti and Reiner. He spent the summers of 1943–5 at a sanatorium by the Saranac Lake, again at the expense of ASCAP, and his health began to improve, so that in his first summer there he was able to complete the Koussevitzky commission, the Concerto for Orchestra. In November 1943 he met Yehudi Menuhin, who asked him to write a violin sonata, and for the winter months he was sent by ASCAP to Ashville, North Carolina, where he finished the Sonata for solo violin in March 1944.

Despite arrangements made with Columbia University by Bátor for Bartók to resume his appointment (and to be paid from funds collected by Szigeti from musical organizations and friends), his doctor advised more rest and he never took up the offer. He continued work on his Romanian collection, completing a third volume of some 1700 texts, and he was able to be present at the first performances of the Sonata for solo violin (Menuhin, New York, 26 November 1944) and the Concerto for Orchestra (Koussevitzky, Boston, 1 December 1944). His feelings and circumstances were now greatly improved; at Christmas he could write to his pupil Wilhelmine Creel: 'Our modest future is secured for the coming three years'. Yet he did not feel settled. With the USA's entry into the war his connections with Hungary had been severed, and when the war in Europe ended he was informed that, because of transport and official difficulties, he could not think of returning home. He wrote to Eugen Zádor: 'Heaven knows how many years it will be before Hungary can pull herself together in some measure (if at all). And yet I, too, would like to return, for good ...' (1 July 1945). Nevertheless, in 1945 he worked on three new compositions. The Third Piano Concerto, intended for his wife, was completed but for the orchestration of the final few bars; the Viola Concerto, commissioned by Primrose, was left in sketch (both works were completed by Tibor Serly). Bartók also planned a seventh quartet, but in September 1945 his health suddenly took a turn for the worse. From Saranac Lake he travelled back to New York, was taken to West Side Hospital, and there died on 26 September. His widow moved to Budapest in 1946 and continued to play in two-piano recitals of Bartók's works and to record them. She died in Budapest on 21 November 1982. Her son Péter Bartók, a sound engineer and editor for Bartók Records in New York, is an American citizen. Béla Bartók jr, an engineer, has remained in Budapest and is the author of several documentary studies of his father.

CHAPTER SIX

Early works, 1889–1907

It was decisive to the formation of Bartók's style that he was not a prodigy composer and was not treated as such; a career as a piano virtuoso seemed to promise more. He had, it is true, written some 70 works by the age of 22, but these are hardly original; rather they show highly skilled imitation, assimilation by self-education or exercise in instrumental forms. The works of this early period show a series of sudden advances under the influences of new teachers, so it is possible to classify Bartók's juvenile output into five distinct periods (this categorization was established by Dille in his thematic catalogue).

The first period works are those of Bartók's first opus numbering (31 pieces, autumn 1889 to spring 1894). They were written during the years of migration from one country town to another, when Bartók had two notable piano teachers (Kersch and Burger) but no training in theory. Most of the pieces are dances for the piano – polkas, waltzes, mazurkas and Ländler – and other short items; even the 573 bars of *The Course of the Danube* are divided into several small movements. Bartók's model here was local utility music.

A new opus numbering marks the beginning of the second period, when the family was settled in Pozsony and Bartók was having systematic piano lessons from László Erkel (July 1894 to December 1896). He had still received no instruction in harmony, but was encouraged by Erkel to observe the examples of his new piano repertory, as may be discerned from the Mozartian F major Piano Sonata or the distinctly Beethovenian sonatas for piano in G minor and for violin and piano in C minor. The works of 1896, among them two string quartets, have unfortunately been lost.

The compositions of Bartók's third early period (1897 to summer 1899) show his rapid acquisition of a surprisingly mature technique and fluent style, thanks to the methodical German character of Anton Hyrtl and to his harmony and theory teaching, and also to Bartók's own fast-growing appreciation of Classical and Romantic music. He continued to follow Beethoven but also looked to Chopin, Mendelssohn and, above all, Brahms, whose rhythm, thematic processes and chamber textures left their mark on the A major Violin Sonata and the C minor Piano Quartet. Bartók's first vocal essays, some songs to German words, are pale imitations, and it was becoming obvious that he was a born instrumental composer, with an instinctive sense of form and balance. He was at his best in scherzo movements, where metric and rhythmic complexity is achieved by means of Brahmsian hemiolas, as in the Piano Quartet and the F major String Quartet. Hungarian elements – apart from certain heroic, dotted-rhythm themes of Brahmsian character – are absent: Bartók was on his way to becoming a composer in the cosmopolitan-German manner.

With his move to Budapest, Bartók began regular academic studies in composition and the piano, and his view of musical values and problems suddenly broadened; his music underwent another change, and a fourth period can be dated from autumn 1899 to spring 1904. The first half of this period, however, saw little original work, apart from the *Liebeslieder* and a few piano pieces with Schumann-like ciphers inspired by his love for Felicie Fábián. His teacher Koessler, a Brahms enthusiast, dampened his creative impetus, and he was made to write exercises instead of a piano quintet. (The distinction between deliberately fragmentary exercises and compositional sketches, or between complete dance forms written for practice in orchestration, and genuine compositions, is not always properly made in Dille's catalogue, see for example DD 58–61, 65.) 'From this stagnation', he later wrote, 'I was aroused as by a flash of lightning by the first Budapest performance [February 1902] of *Also sprach Zarathustra*'. He studied Strauss's symphonic style and motivic fabric, and he gained a closer acquaintance with the work of Liszt.

According to Bartók's autobiography of 1918, at that time he became conscious of his duty to shape a specifically Hungarian style, following the general feeling of nationalism and revolt against the Habsburg hegemony.

He wrote his first Hungarian settings, the Four Pósa Songs, in 1902, choosing the *népies műdal*, or popular song style. Other works of 1902–3 still represented a transition, during which the combination of Straussian harmony and expressiveness with 19th-century-type Hungarian motifs and rhythms, and of leitmotif technique with classical formal development, led to works of a certain complication and structural weakness. Examples include the unfinished Symphony, of which the scherzo served as Bartók's diploma composition, and the Violin Sonata, with its disproportionate finale.

Early works

At the same time, in both small and large pieces he produced the first compositions that have distinctly Bartókian features: the symphonic poem *Kossuth*, the scherzo (no.4) of the Four Piano Pieces and the Piano Quintet (written in response to Dohnányi's celebrated Piano Quintet op.1 of 1895). This last work, despite its formal redundancy and eclectic style, is his earliest significant chamber composition.

The works written between summer 1904 and 1907 may be regarded as representing a fifth early period (even after the establishment of his mature style, he was able to identify with four large-scale late Romantic compositions of this period, retrospectively numbering them opp.1–4 in his third and final classification of 1908).

At this time Bartók had begun collecting folk music in Hungarian villages, but his stylistically divergent and exploratory arrangements of folk music, the first being in the volume produced with Kodály in 1906, were secondary to the creation of Hungarian concert music, 'Hungarian' in the late 19th-century sense. Bartók's opp.1–4 constitute a typically Hungarian, fin-de-siècle culmination to the symphonic tradition of Erkel, Mosonyi and Liszt. There is enduring value in the post-Lisztian but individual Rhapsody op.1 for piano and orchestra; in the First Orchestral Suite op.3, with its pseudo-Hungarian 'Meistersinger' polyphony and its signs of a linking monothematic organization; and in the Second Orchestral Suite op.4, which suggests a potential Hungarian impressionism or serenade-like lyricism. These works are of further interest in their adumbration of Bartók's mature compositional principles, formal schemes, thematic types, scoring and harmonic effects. There is the outline of a five-part palindromic form in the First Suite, for example, and the Second contains such characteristic features as frequent metric change, modal cadences, and a closing chord of F♯–A–C♯–E. Moreover, the Scherzo op.2 'for orchestra and piano', though fairly eclectic in its material, is very much a key work, with its grotesque, burlesque *ma poco variato* reprise which caricatures Romantic substance and involves the sorts of thematic and instrumental effects that were to become the modern, Bartókian elements in his mature style of the next decade. His last important late Romantic score was the Violin Concerto Sz 36, a piece more influenced by his new contacts with contemporary music (in 1907, for example, he acquired a number of Reger scores) and by his development of Lisztian thematic metamorphosis than by the stimulus of folk music.

CHAPTER SEVEN

Establishment of mature style, 1908–11

In considering the appearance of Bartók's original musical language about 1907–8 it is impossible to overstate the special nature of his dependence on peasant music – something that many contemporary aestheticians, among them Adorno, failed to understand, so that they took a fundamentally distorted view of Bartók's development. His dependence on folk music was quite individual: to all intents and purposes he felt more confident in his innovations when they seemed to him to have some justification in the various kinds of folk music that his research was making more widely and scientifically appreciated. Sometimes he was directly stimulated to arrive at new ideas by the peculiarities of peasant music, its microtonality, its freedom from major and minor scales; sometimes the ideas came from his growing awareness of new art music (that of Debussy, and later Stravinsky), but could be 'legitimized' and developed in a completely individual way only when he found a counterpart in folk music. Whichever the case, just as Schoenberg justified his innovations by appeals to the great German tradition, Bartók had a decisive, almost ideological belief in the example provided by the 'natural phenomenon' of folk music. As he himself stressed in one of his late American lectures (see Vinton,

6. Bartók collecting songs from Slovak peasants in the village of Darazs, 1907

ed., *JAMS*, 1966; and *Béla Bartók Essays*, ed. B. Suchoff, 1976, p.363):

the start for the creation of the new Hungarian art music was given first by a thorough knowledge of the devices of old and contemporary Western art music (for the technique of composition); and second by this newly discovered musical rural material of incomparable beauty and perfection (this for the spirit of our works to be created).

Scores of aspects could be distinguished and quoted by which this [peasant music] material exerted its influence on us; for instance: tonal influence, melodic influence, rhythmic influence, and even structural influence.

In 1907–8 Bartók also found stimulus in contemporary art music, primarily in its tonal aspects and its redefinition of consonance and dissonance; in the loosening of tonality he went through an evolutionary process paralleling those of his contemporaries. Influenced perhaps by Reger, whose works he perused assiduously in autumn 1907, he experimented with post-*Tristan* chromatic melody and with the intensive development of motifs, particularly those containing declamatory augmented or diminished intervals (see the first of the *Two Portraits*, the opening movement of the First Quartet and the first *Elegy*). While engaged in these endeavours he developed the leitmotif D–F♯–A–C♯, which in 1907–8 was the motto for Stefi Geyer; later it acquired a more general significance, denoting longing or menace, as in Judith's cries in *Bluebeard's Castle* or the calls of the father in the *Cantata profana*. He also explored superimposed 3rds (in the fourth *Sketch*), pieces built from one or two chord structures (in the third and fourth of the *Négy siratóének*, 'Four dirges'), alternating chords on the black and white keys (in the second *Burlesque*) or the black and white contrast of melody and accompaniment (in the seventh Bagatelle). At the same time, a piece such as the second *Elegy* displays a hyper-Romantic bravura prompted by Liszt's transcriptions, though coupled with strict motivic organization.

However, these developments in line with new art music proved less decisive than the influence of peasant music, for in the latter Bartók found not only the absence of the major–minor system but new kinds of tonal structure from which he could form a 'tonality'. It was of particular significance when, travelling in Transylvania in summer 1907 and discovering numerous pentatonic tunes on A sung there, he noted down a simple scale (ex.1) for his own use. Partly influenced by further experience of live folk music and partly as a result of speculative work sparked off by study of the

Ex.1   Ex.2   Ex.3

Debussy scores in his possession in 1907–08, Bartók used this scale as a starting-point. He realized at once that the notes of the anhemitonic pentatonic scale sound consonant when played together, regardless of which is the bass note (ex.2). From this he found that a mixture of complete or incomplete pentatonic clusters and inflected pentatonic chords (as dissonances) could provide a marvellously homogeneous harmonic system (see the fourth and fifth Bagatelles). It was from the pentatonic scale, or rather from certain Hungarian folksongs rich in melodic steps of a 4th, that he derived chords of 4ths (ex.3) containing both perfect and imperfect intervals, as in the Bagatelles. (In 1908 the only Schoenberg work he knew was the First Quartet, and so the Western example of such 4ths must have been Debussy.)

At the time of the Bagatelles Bartók's harmonic vocabulary was still eclectic, including major and minor triads and tetrads constructed of 3rds, chords in 4ths etc; but his experiments already included chords made up of typically pentatonic intervals (integral multiples of two, three and five semitones) and these began to predominate. The most characteristic of these 'pentatonic chords' quickly became the Bartókian major–minor chord. Although Bartók may have noticed this chord in Liszt's music, he remarked: 'It is very interesting to note that we can observe the simultaneous use of major and minor third even in instrumental folk music' (see Suchoff, ibid, 369). Unlike Stravinsky, Bartók mostly had the minor 3rd uppermost, and after a time the chord appeared consistently only in its first inversion (ex.4). Lendvai has called this pentatonic chord the 'alpha chord' (ex.5) and has classified various derivatives (ex.6).

Ex.4   Ex.5   Ex.6

The second important discovery Bartók made in folk music in 1907–8 was the old modal scales; they offered the possibility of eliminating leading-note dominant-tonic cadences and of establishing a certain amount of

41

equality among chords based on different degrees of the scale. This was first put to the test in the Three Hungarian Folksongs from the Csík District. Furthermore, modality led Bartók to a loose tonality in which a melody (or a formal section, a movement or a whole composition) can begin and end with a clear statement of a mode but can alter and diversify in modality along the way. Together with Dorian and Mixolydian modes on the same note, a combination of Lydian with Phrygian proved particularly useful, especially in Bartók's rapidly developing notion of what he called 'polymodal chromaticism': 'As a result of superimposing a Lydian and a Phrygian pentachord with a common fundamental tone, we get a diatonic pentachord filled out with all the possible flat and sharp degrees (quoted in Suchoff, ibid, 367). The overlaying of different modes was used experimentally in the first Bagatelle (14 April 1908), which bears a key signature of four sharps in the upper staff and four flats in the lower. At the end of his life Bartók attempted to distinguish such procedures from the conception of bitonality then in fashion, noting that his was 'simply a Phrygian-coloured C major', as the second *Sketch* was 'indisputably a pure C major' (see Suchoff, ibid, 433).

From 1908 onwards Bartók found ideas for the successive and simultaneous juxtaposition of different modes in Romanian folk music; significant in this were his first original compositions 'in Romanian style' (the sixth *Sketch*, the *Két román tánc*, 'Two Romanian dances', and the second of the *Two Pictures*). After Slovak melodies with their Lydian 4ths, Romanian tunes offered examples of a lower Lydian pentachord with a Mixolydian upper part, which together produce

7. *Bartók's first draft (14 April 1908) for the first of the Fourteen*
Bagatelles

the 'acoustic scale' (ex.7). Lendvai rightly saw in this, and in its melodic and chordal derivatives, the most characteristic basis of what he termed 'Bartókian diatonicism'. Earlier, in the seventh *Sketch*, Bartók had exploited the fact that a single step separates the acoustic scale, which produces the most static type of tonality, from the atonal whole-tone scale; he may have found both in Debussy. The two complementary whole-tone clusters (Lendvai's 'omega chords', ex.8) proliferate as alternating chords in Bartók's music of the early 1920s

Ex.7 Acoustic scale   Ex.8

(see the second movement of the Second Violin Sonata, from figure 22) and become truly constructive in the 1930s (see the first movement of the Music for Strings, Percussion and Celesta). The acoustic scale itself quickly became associated for Bartók with scenes of nature (the opening of *The Wooden Prince*, the garden scene in *Bluebeard's Castle* and the chorus *Tavasz*, 'Spring'), rural images (the opening themes of the Divertimento and *Contrasts*) or endings glorifying freedom (the *Cantata profana* and the Music for Strings, Percussion and Celesta).

Another structure of key importance in Bartók's tonality and chord formation, that of scales without a fixed final note and containing repeating fragments within the octave, also belongs among the basic elements endorsed by folk music, though he became properly aware of it only after his collecting tour in the Biskra area in 1913; he used such scales in, for example, the Piano Suite op.14 and the Second Quartet. In Lendvai's apposite terminology based on semitone structure, these scales may be called 'model 1:5', 'model 1:3' and 'model 1:2' (see ex.9), the last being identical with Stravinsky's 'octatonic scale' and with Messiaen's second mode of limited transposition. In recent analyses (Treitler, Antokoletz) chords of the 'model 1:5' structure are called 'Z-chords'.

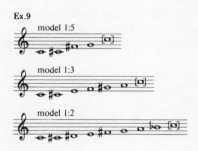

Ex.9

Since Bartók himself clearly differentiated between 'diatonic' and 'chromatic' versions of the same themes, indeed regarded them as opposites (see his preface to the pocket score of Music for Strings, Percussion and Celesta, the Fourth Quartet, etc), the search for a special tonal system and the semantic explanation of it is justified. Ernő Lendvai's view is that Bartók's 'chromaticism' is distinguished by intervals based on the golden section, or, in numerical terms, on the Fibonacci series 2–3–5–8–13–etc. Thus the pentatonic scale, the alpha chords and the modal scales all belong to this system. (According to Bartók's own, different classification, his first 'chromatic' melody was the opening theme of the Dance Suite; see Suchoff, ibid, 379). Lendvai also

45

connected all these things through what he called the 'axis system', which affirms the presence of tonic–subdominant–dominant chord functions, or rather tonal relationships, in Bartók's music. If functional harmony in the classical sense is not typical of the small-scale harmonic progressions of any of Bartók's mature music, Lendvai's axis system, in which the three functions are held by the three diminished 7th groupings (see ex.10), is undeniably present. In essence, it functions as an organizing principle of tonal levels or centres in movements or movement sections, a structure in which all 12 notes are related to one centre. Bartókian

Ex.10

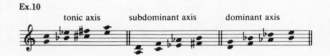

tonic axis   subdominant axis   dominant axis

'diatonicism', according to Lendvai, is virtually the opposite of his chromaticism; the diatonic system includes the acoustic scale, chains of open 4ths, chords of accumulated 3rds or open 5ths, augmented triads and so on. (However, the limits of a universal tonal system like that of Lendvai are obvious: technical procedures can too easily become identified with the alleged poetic meaning of the music, and there is a tendency to discover supposed dialectic structures of a thesis–antithesis nature. Nevertheless, acknowledged analytical trends with a narrower scope, for example those in search of 12-note procedures (by Mason, Babbitt, etc), symmetrical formations in pitch organization (by Treitler, Perle, Antokoletz, etc) or Schenkerian explanations (Waldbauer), even with a sophisticated new terminology

seldom go beyond the exploitation of technical procedures in isolated works and usually ignore Bartók's own
musical experiences, development, and rhythmic and
thematic originality.)

Around 1908 Bartók's dependence on folk music
became still more marked in the areas of theme formation, melody and rhythm. He continued to use several
varieties of instrumental theme quite unlike anything in
peasant music, such as post-Wagnerian unending
melody (in the first of the *Two Portraits* and the First
Quartet) or a more grotesque development of the
Lisztian 'valse diabolique' (in the 14th Bagatelle, the
second of the *Two Portraits* and the third *Burlesque*);
but his thematic creativity became manifestly shaped by
study of folk music, and in various ways. He wrote
themes sounding like imitation folksongs (e.g. the fifth of
the Ten Easy Piano Pieces, which mimics the parlando
rubato vocal melody and the *giusto* instrumental style of
Transylvanian music), themes based on particular types
of folk rhythm (e.g. the tenth of the same cycle, with its
swineherd's rhythm), themes using the characteristic
four-line strophic form of Hungarian and Slovak folk
music but lacking any folk flavour (e.g. no.2), and
themes that are refined abstractions of folk music's linear structure, tonality and rhythmic organization but
that do not evoke any specific folk dialect, sounding
simply 'Bartókian' (e.g. no.1). Generally speaking, certain principles of folk music became in these early years
virtually automatic in his thematic invention, for
example the tonal or modal unity and the simplicity of
melodic sections, the isorhythmic conformity of sections, the kaleidoscopically varied repetitions of tiny
motifs within the theme, the repetitions and interpola-

tions of bars as in Romanian instrumental folk music, and the extension of intervals of a theme or motif when repeated in a different scale or mode. All of these can be observed in one of Bartók's outstanding works of the period, the *Allegro barbaro*.

In formal terms Bartók worked to a large extent within the bounds of tradition during the first years of his maturity. For smaller works he chose the framework of simple or multiple ternary form; for larger ones he took the models of sonata and sonata rondo, evidently on the basis of an intensive study of Beethoven (see the First Quartet). However, other schemes did appear in very short movements: folksong-like forms of one or several verses, structures organized round an ostinato, and what might be termed 'chain forms', in which the recurrent recapitulations that provide balance are of secondary importance to the progressive and successive variation of a fragment. This last type was suggested by Romanian instrumental music and is exemplified by the second Romanian Dance. (It is worth considering, too, whether the homophonic textures and sectional thinking that predominated in Bartók's music throughout his life did not arise from the early, determinative influence of folk music.)

Many of Bartók's early multi-movement works were clearly based on models, for example Beethoven (Bagatelles), Schumann (*Für Kinder*) and Debussy (*Esquisses, Deux images*), as the titles alone indicate. He put into such cycles both original works and folk music arrangements, and as a performer did not always play them as integral works. True formal cohesion occurred at this stage only in certain two-movement compositions such as the Violin Concerto Sz 36 (the portrait of a girl and that of a violinist), the *Two Portraits* ('ideal' and 'grotesque') and the *Two Pictures* (nature and village life). In his later works Bartók did not remain satisfied with this Lisztian conception, but he was preoccupied from first to last with contrast structures made up of two movements or built from two species of material, or at least with diametrical oppositions between beginning and close. But by then the order of events and the aesthetic justification were to be directed not by antithesis but by a Beethovenian dramaturgy from conflict to triumph.

CHAPTER EIGHT

Years of extension, 1911–24

The one-act opera *Bluebeard's Castle* (1911) constitutes
not so much a new departure as a consummation on a
higher level. After a period in which Bartók's originality
had been most clearly displayed in shorter piano pieces
(and in which he clearly had had problems with form
and balance in orchestral works), the opera stands as a
masterpiece in a complex genre. It is a remarkable instance of dramatic significance achieved through musical means: for example the contrasting of pentatonic,
folklike material for Bluebeard with chromatic,
Romantic music for Judith; the tonal structure – pentatonic F♯ representing night at the beginning and end,
with a flood of light in C major in the middle as the
fifth door is opened; the subtle interweaving of symbolic
or allusive motifs (the blood motif, the reference to the
St Matthew Passion etc). *Bluebeard's Castle* also introduced a new style of Hungarian recitative, based on
peasant music and having an affinity with Debussy's
musical recitation while being, as Bartók emphasized,
'the sharpest possible contrast to the Schönbergian
treatment of vocal parts' (quoted in Suchoff, ibid, 386).

In formal terms the opera was decisively important,
containing two ideas central to Bartók's mature structure: one was the idea of palindromic form, which was
to recur more explicitly and in thematic terms in *The
Wooden Prince*, and was to be, from the Fourth Quartet

onwards, his principal architectural device in instrumental music; the other was a sort of rondo striving towards a large coda. In the opera a series of static tableaux interrupts the flexible opening music, which is unfolded only at the end. *The Miraculous Mandarin*, with its two dance episodes inserted into the sequence of three seduction attempts, its central waltz and chase, and its threefold murder and grand ending, is another example of this form in spite of its symmetry. A more clearcut case is the Dance Suite, in which short movements linked by ritornellos lead to a big finale; but the Sixth Quartet exhibits this quasi-rondo form at its culmination.

All three of Bartók's stage works are concerned with the relationship between man and woman, each proposing a different solution which determines the musical form. The ballad-style tragedy of *Bluebeard's Castle* shows the opening and closure of the man's castle, his soul. *The Wooden Prince* has a story of folktale optimism: the man forgives, nature provides the release, and the work returns to the lush atmosphere of its beginning. *The Miraculous Mandarin* is again tragic, but cathartic: the primitive (i.e. natural) man's desire and the corrupt (i.e. civilized) woman's purification in love are consummated at the moment of death. It is noteworthy that Bartók's two song cycles of 1916, opp.15 and 16, written at a time when his private life was in crisis over the Klára Gombossy episode, are also organized round the theme of love and death, and that his two main instrumental works of the same period, the Piano Suite op.14 and the Second Quartet, are related formally: gradually accelerating movements reach an ecstatically fast tempo, and are followed by tragic slow finales. However, the Four Orchestral Pieces (1912), which close with a funeral march, indicate that large-scale forms culminating in tragedy cannot be related to a single biographical incident. Bartók's outlook at that time was considerably affected by all manner of setbacks and crises: the failure of *Bluebeard's Castle* in the competition, his depression brought about by the outbreak of World War I, and so on.

In terms of the further evolution of Bartók's style, *The Wooden Prince* worked as a kind of catalyst, for in it he brought together the whole stock of late Romantic devices accumulated since the Scherzo op.2, assembling them in what is almost a persiflage and including genuine innovations only in the grotesque puppet music, which exemplifies his '*allegro barbaro* style'. The scoring is slightly harsh and sometimes bizarre; folksong-like themes, for instance, are given rather oddly to the saxophone, and the xylophone is prominent (though both these details were smoothed out in the course of revision). Furthermore, the ballet has undeniable formal weaknesses, which Bartók was aware of and which necessitated cuts in the theatre, some of them authorized by him. After this a change of direction was unavoidable, and it came on two fronts. First, while finishing the ballet he was already striving for much more economical, concise and transparent effects as a counterpoise to the extravagance of some sections (this comes out most strongly in his chamber and instrumental music; the austere effects of the Piano Suite op.14 and the splendidly sparing use of intervals in the Lento of the Second Quartet point towards the future). Second, he began to take a more systematic and disciplined hold of the formal dramatic confrontation between material

8. Autograph MS of the first page of the piano score of 'Bluebeard's Castle', composed 1911; the German translation (unpublished) is in Mrs Emma Kodály's handwriting

inspired by folk music and that of a more personal, more expressive and Romantic type.

A new formal control was also entering Bartók's cycles of arrangements from folk music at this time. He was beginning to make his sets ethnically homogeneous and to organize them into 'classical' three- or four-movement patterns to be performed as uninterrupted wholes. Examples include the 20 *Román kolinda-dallamok* ('Romanian Christmas carols'), the 15 Hungarian Peasant Songs arranged in four sections, each distinguished by a particular folk style, and the eight Improvisations on Hungarian Peasant Songs, which comprise four tonal blocks put together according to a certain succession of intervals. All of these are for the piano, and all date from 1914–18. A later vocal work, the *Village Scenes* (1924), exhibits a still firmer unity: the songs are all taken from one Slovakian county, and they make up a balanced musical form recording the customs of peasant life.

The outstanding composition of this period and the greatest of Bartók's dramatic works is *The Miraculous Mandarin*, in which the choreography is virtually mapped out in the strongly defined gestures of the music. If it had been performed at the time of its completion in piano score (1918–19), it might have produced a sensation to rival that of *The Rite of Spring*, which, it should be noted, Bartók knew only in piano score (and at that time the only atonal works of Schoenberg's he knew were opp.11 and 19 and the four-hand piano score of op.16). The orchestration of the ballet, in the absence of a promise of production, took some time, and Bartók prepared himself by orchestrating the Four Orchestral Pieces and the Dance Suite – studies for the exceptionally vigorous and inventive instrumentation of *The Miraculous Mandarin*. One point of particular interest is the use of quarter-tones to define a slow string quasi-glissando. In the 1937–8 Violin Concerto and the Sixth Quartet Bartók was to use quarter-tones in a more exposed fashion, and longer quarter-tone passages of structural importance are to be found in the original version of the fourth movement of the Sonata for solo violin.

It is no accident that *The Miraculous Mandarin* and a few contemporary works, such as the piano studies and the two sonatas for violin and piano, are described as belonging to Bartók's 'expressionist period', thus emphasizing that at this time he came closest to the aspirations of the Second Viennese School. Free 12-note collections are common in these pieces; the tonal centre sometimes remains obscure for several bars; textures are dense; and the rhythm is often extremely complex. Referring to this period in his American lectures of winter 1927–8, Bartók himself said: 'There was a time when I thought I was approaching a species of twelve-tone music. Yet even in works of that period the absolute tonal foundation is unmistakable' (see Suchoff, ibid, 388). A good example of a 'tonal foundation' organized with typical Bartókian logic, despite apparent looseness, is provided by the Violin Sonata no.1 (1921), which he described as 'in C♯'. The first movement begins with a seemingly polytonal opposition between C♯ and C, from which A minor emerges to temporary dominance at the end of the movement, reaching this position through textural means as the violin melody comes to the fore (see ex.11*a*). The second movement is less ambiguously centred on C (minor), even when two more levels are added simultaneously to the main level of A–C–F♯–(E♭) (see ex.11*b*). The third movement has at its core material of an instrumental folkdance character, and is therefore more vigorously tonal, remaining so where the violin's main theme in 'heptatonia secunda' on B is in dissonant opposition with the C♯ tonality of the piano. ('Heptatonia secunda' is the term introduced by Lajos Bárdos to designate the scale whose intervals are successively two, one, two, one, two, two and two semitones; its various transpositions and their derivatives are related to certain features of Romanian and Arab folk music.) The sonata's final chord, with its insistent acoustic 7th and major and minor 3rds, has a definite, pure Bartókian C♯ tonality (see ex.11*c*).

The two splendid violin sonatas were decisively important for the evolution of Bartók's style both aesthetically and formally. In the Sonata no.1 he introduced the idea of following a sequence of subjective movements with a joyous Beethovenian dance fantasia, a sort of collective emancipation; in this case the themes are for the most part primitive, some of them Arabic in flavour, and reminiscent of Romanian and Ruthenian folkdances for violin. The general idea became the basis for the whole form of the Dance Suite of 1923, in which the themes are Hungarian, Romanian and Arab, mostly having an extremely primitive character (a movement omitted by Bartók was somewhat Slovak in nature). In later works, too, particularly the Divertimento and the Concerto for Orchestra, Bartók returned to this scheme. This new form also, of course, has aesthetic implications. Since Bartók no longer felt the simple opposition

Ex.11 Sonata no.1 for violin and piano
(a) 1st movt

(b) 2nd movt

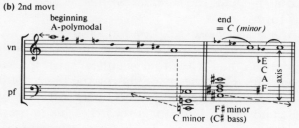

(c) 3rd movt

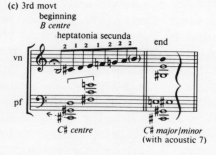

of sensitive lyrical to rustic dance material or movements to be convincing, he sought some kind of formal synthesis and experimented with a pattern that can be likened to the Beethovenian sonata rondo with its long coda or, more closely, to the form of Liszt's B minor Sonata; but in essence it is original. Its components are the exposition and development of a lyrical movement,

the exposition and elaboration of a rustic dance movement, a reprise of the lyrical material in a highly modified form, and a similarly altered reprise of the rustic
dance. This continuous form, which can be represented
as ABa^1b^1, was realized in its purest form in the Third
Quartet (1927), but it had its origins in the finale of the
First Violin Sonata. The form of the Second Sonata is
more complex, since the A material reappears trans-
figured at the end; the *Cantata profana* is also a variant
of this form.

Bartók's journeys to London and Paris around
1922–3 and his participation in the international forum
of new music had a twofold significance in his artistic
development. First, he made up for the interruption of
the war and discovered how Schoenberg, Stravinsky and
the rest had progressed (for a record of his impressions
and disappointments, see his letters to Heseltine published in *Documenta Bartókiana*, v (1977); an odd proof
of his curiosity is that he copied several piano pieces
by Lord Berners). At the same time he was able to take
stock of what his fellow composers and the public
found most vital in his own piano and chamber music.
This was perhaps more influential than the considerations enumerated earlier in bringing to an end, by
and large, the increasing opulence of his music. From
this point he was consciously to pursue economy, to
purge his style as far as possible of the 'common language' of his generation, and to reinforce what was his
alone – or what could be derived from folk music.

CHAPTER NINE

Classical middle period, 1926–37

After the Dance Suite (1923) came a pause of two and a
half years, often called a barren period, though it was
in fact the prelude to a new period. This is shown in
the piano part of the *Village Scenes* (1924), the piano
transcription of the Dance Suite (1925) – an arrangement made at the publisher's request of a work that was
having a string of international successes – and the
orchestration of the *Village Scenes* (1926), completed
under the direct influence of Stravinsky, who had just
played his Concerto for piano and wind in Budapest.
Summer 1926 was one of the most radical turning-
points in Bartók's career. Answering a question in the
first significant study of his style, by von der Nüll
(1930), as to why his music was so lacking in counterpoint, Bartók emphasized, with reference to 1926: 'In
recent years I have considerably occupied myself with
music before Bach, and I believe that traces of this are to
be noticed in the [First] Piano Concerto and the Nine
Little Piano Pieces'. (Indeed, from autumn 1926 he frequently played his own transcriptions of Della Ciaia,
Michelangelo Rossi and others, the majority of them
taken from Torchi's *L'arte musicale in Italia*.) Although
there are probably textural or even spontaneous motivic
references to his Baroque repertory, the mark of the
decisive change is not so much the relative increase in
counterpoint (as in the free fugato in the second move-

59

ment of the First Piano Concerto, the abundant tiny imitative passages in its fast movements, the two-part invention forms in the Nine Little Piano Pieces etc) as the distinctively Bartókian neo-classicism evident in the proportions of his compositions, their forms and, above all, their purified textures.

Although certain striking Stravinskian features appear in Bartók's music at this time (e.g. in the development sections of the opening movements of the first two piano concertos), Bartók remained quite apart from the neo-classical endeavours of many others. He did not write movements or works after models from the past; he chose this moment to introduce a series of decidedly progressive, even experimental, instrumental devices into his work (e.g. his treatment of the piano as a percussion instrument, and in particular his use of clusters, following Cowell, in the First Piano Concerto and the fourth piece of *Out of Doors*, both of 1926; and the 'Bartók pizzicato' first used in the Fourth Quartet of 1928); and, far from eliminating folklike thematic material from his original compositions, he strengthened this aspect and sought still more consistent solutions to the formal problems raised (see Somfai, *Tizennyolc Bartók-tanulmány*, 1981). This is exemplified in the two-movement structure based on 'abstract' and 'concrete' forms of the same material in 'Preludio – All'ungherese', the last of the Nine Little Piano Pieces; in the use of folklike themes as second subjects in sonata forms, as in the First Piano Concerto; and in the closing of an exposition with a cheerful folk theme, as in the first movement of the Piano Sonata.

The two great piano concertos, nos.1 and 2, were undoubtedly written for Bartók himself as an international virtuoso, and to some extent they were designed specifically to fit his gifts. Only in exceptional cases did he allow other pianists to give a national première of either work. The concertos contain passages where the performer determines the character with the authority of a Romantic keyboard lion, but these are contained in extremely tightly organized symphonic forms. This duality led to one of the most notable of Bartók's revisions of sonata form: in the first movement of the Second Concerto, an orthodox sonata structure is almost interrupted by the ostinatos, outbursts and cadenza of the impulsive piano, and the form is held together by a brass motto placed round the sonata like a hoop, following its own circuitous yet logical path of mutation and imitative iteration. This motto, a quotation from Stravinsky in its original form, becomes a splendid Hungarian theme, in both melody and rhythm, in the retrograde inversion before the cadenza.

By contrast with these great 'ego' works, the middle string quartets were rapidly accepted as masterpieces; the scoring is brilliantly idiomatic, and also definitive, so that to a much greater extent than in the piano concertos the interpretation is written in. (It is worth noting that before he drafted the Fourth Quartet Bartók was acquainted with Berg's Lyric Suite, so rich in similar effects.) After the condensed and continuous ABa^1b^1 form of the Third Quartet, mentioned above, Bartók sought a more spectacular structure that would display its organization more clearly. This he found in the palindrome, which defines the fine classical symmetry of the Fourth and Fifth Quartets, their sonata-like balance between movements of 'scherzo' and 'nocturne' type, and even the organization of thematic blocks within movements. Palindromic form had been introduced, as has been shown, in the stage works; it was introduced into instrumental music in the second and third of the Three Rondos (1927), and it became elaborated further in the five-movement form of the Fourth Quartet (1928). Bartók described the form in his analytical introduction to the miniature score of this last work (1930), and in a similar essay on the Music for Strings, Percussion and Celesta, published in 1937, he used the term 'Brückenform' ('bridge form') to describe the palindromic structure of the third movement.

The essence of this form, which customarily follows the five-component scheme $ABCB^1A^1$, is that the fourth and fifth sections are not just variations on the second and first but are recast to produce something aesthetically conclusive. For this structure, containing so much quasi-geometrical symmetry, is not static: it does not return to its origins but progresses towards a cathartic outcome. Thus the nature of the central section is all-important. In the metaphor of the *Cantata profana* it matters enormously what happens at the 'crossing of the bridge', at the point of the young men's 'transformation into stags'. The Fourth Quartet contains at its centre a Hungarian adagio with a nocturnal monologue; the Second Piano Concerto has a toccata–scherzo of visionary flashes which brings to life a frightening natural scene dispelled at the call of the horn; and in the Fifth Quartet it is a wild dance in Bulgarian rhythm.

One type of the five-part form has two scherzos surrounding a slow movement (as in the Fourth Quartet); the other has two slow movements around a scherzo (as in the Fifth Quartet and the Second Piano Concerto, where the BCB^1 segment is the second movement). Later Bartók wrote three-part forms in which the *BCB*[1] section lacks the contrast of scherzo and adagio (e.g. the Divertimento), or in which it is replaced by an asymmetrical structure (e.g. the large-scale set of variations that forms the central movement in the 1937–8 Violin Concerto). In his palindromic forms the outer movements are sometimes related in such a way that a new theme in the finale provides a framework for the reworking of first-movement material, as in the Second Piano Concerto.

Gradually Bartók extended the palindromic principle to the sonata structures of opening movements, so that the recapitulation brings back the themes in retrograde order and in inversion, as in the Fifth Quartet. This principle, however, is never strictly implemented; there are 'irregular' deviations which may be said to signify creative strength. In the first movement of the Violin Concerto (no.2), for instance, the reprise of the subject is purely fragmentary; the theme virtually 'works itself out' after the cadenza, but now in its basic form and to great expressive effect. The elaboration and thorough implementation of palindromic form thus led Bartók to reconsider and in certain aspects to simplify his style, but at the same time to intensify contrasts between extremes. Examples of this include the dialogue between open 5ths and unison melody in the Adagio of the Second Piano Concerto, and the total opposition of playing techniques between the second and fourth movements of the Fourth Quartet. Instrumental themes and motifs henceforward became simpler, dependent on fewer intervals, rhythmic values or other units, so that the idea might remain recognizable in inversion and various modified forms. In British and American analytical literature this mature but transitory state of his instrumental style has often been considered a typically Bartókian idiom.

The fundamental individuality of Bartók's mature style – its newness and perhaps its restrictedness too – lies to some extent precisely in this network of inventively simplified structures carried to logical extremes. Several techniques became almost reflex actions: the widening and narrowing of intervals, the lengthening and shortening of numbers of bars or beats, the constant alteration of a motif, and so on. Sometimes such methods even became mechanical. Just exactly how Bartók calculated each element has been discussed by many writers in analyses of *Mikrokosmos*; the 44 Duos for violins, the folksong cycles and the 27 unaccompanied choral pieces are also to some extent studies in compositional technique. It is well known that Bartók refrained deliberately from Schoenbergian serialism; his 12-note themes, such as that of the Violin Concerto (no.2), are merely playful variegations. Yet it is equally clear that he put together a strict set of rules for his own use, rules which he broke time and again in the relentless reworking of his style.

Apart from palindromic form, which was perhaps Bartók's greatest formal discovery and at the same time the compositional 'game' that he played most frequently, he consciously worked out several other forms of different strengths and aims. He thought very much in terms of genre, making clear distinctions between important works, popular pieces and educational music; and he never abandoned the practice of writing easier pieces to facilitate understanding of his art. For example, between 1931 and 1933 he was at work on the *Magyar képek* ('Hungarian sketches') for orchestra and on several orchestrations of his lighter piano pieces, and later he planned other adaptations, including a string orchestral version of the Fourth Quartet (see Kroó, 1969; Eng. trans., 1970). His series of increasingly refined arrangements of folk music, written to popularize that art, followed a separate path and contained ideas whose qualities have not been recognized sufficiently. In the two violin rhapsodies, for instance, he took Transylvanian dance music, originally for the violin, and wove it into music united by a common peasant function or performing style. Similarly, he organized the 20 Hungarian Folksongs for voice and piano so as to display the historical and stylistic types on which he had drawn. So artful, indeed, are these folksong arrangements, particularly in the originality of their accompaniments, that they may be said to be original songs.

The *Cantata profana* (1930), written as the first part of a trilogy (or possibly tetralogy) of cantatas, occupies a special place in Bartók's output. Superficially the work appears to be one of his most strongly neo-classical in its initial allusion to the *St Matthew Passion*, its quasi-Baroque handling of the chorus, the alternation of arioso and turbae in the second movement and so on. He compiled the text from two variants of a Romanian hunting *colinda* and began to sketch the work with a Romanian text (see László, 1980, pp.213–54). He then made a Hungarian translation – indeed, the score contains not a note of Romanian folk music. In the text he found a symbolic story that could become the grandest exposition of his personal philosophy: he who has regained freedom in nature cannot return to corrupt civilization; he must rather accept solitude, rejection and alienation. The crossing of the bridge, which signifies the transformation of the youths into stags, also stands as a metaphor; and since Bartók carried the symbolism into his music, the *Cantata profana* provides unusually powerful evidence for his musical 'semantics'. As Lendvai observed, the polarization of the 'Passion' beginning against the 'nature' close is defined in the essential material, for the closing acoustic scale of D–C–B–A–G♯–F♯–E–D is the inversion of the opening scale of D–E–F–G–A♭–B♭–C–D, with its insistence on what Lendvai called the golden section intervals.

It is also noteworthy that *Elmúlt időkből* ('From olden times'), a motet-like piece in three movements for unaccompanied male chorus, also has philosophical aspirations, being a eulogy of the peasant way of life. The texts for this and for the 27 partsongs were compiled by Bartók from Hungarian folk literature, and the fact that in these later years he could work only with peasant texts may be due to his inability to link his musical language to the more affected nuances of a poet's text. There was also a practical reason, for from 1934, having given up teaching, he was working on the great edition of Hungarian folksongs and, as part of this, classifying the texts.

The urge to break away from the palindromic stereotype, combined with a determination to preserve its positive features, brought Bartók in 1936 to what is perhaps his most perfectly formed symphonic piece, the Music for Strings, Percussion and Celesta. This much-analysed four-movement structure includes several important innovations: it is monothematic, and the treatment of the theme ranges from quotation to numerous transitional stages of variation; the development runs

9. Béla Bartók, 1936

without interruption through the four movements, so that, for example, the chromatic form of the theme in the first becomes a diatonic version in the last or, again, the melodic convolutions of the first are straightened to generate a simple tune in the last; the work exhibits a new exploitation of space integrated into the formal design, going far beyond the traditional double-choir style, since the centre is also used, and the back and front too, as in the second movement; the chamber-musical, abstract contrapuntal writing is thoroughly and structurally linked with the richly innovatory and ut-terly Bartókian orchestration. A work stylistically akin to the Music for Strings, Percussion and Celesta is the Sonata for two pianos and percussion of the next year, 1937. Here, however, the forms are more traditional; there are three movements, of which the first dominates by its weighty substance. It should be noted that the golden section proportions demonstrated in detail in these and other works by such scholars as Lendvai, Bachmann and Szentkirályi were not consciously planned by Bartók. The study of his sketches, drafts and manuscript corrections (see Somfai, *SM*, 1981) has proved that he never used Fibonacci numbers in calculations nor planned ratios of the golden section (but see also Howat, 1983).

Both the Sonata and the Music for Strings, Percussion and Celesta display the metrical and rhythmic characteristics of Bartók's maturity, and it is these that are perhaps the most original features of his style. Rhythmically he owed at least as much to his observation of practices in folk music as to the European art tradition. His mature instrumental music can be divided into three basic rhythmic types according to his own categorization: parlando rubato rhythm, which includes not only the looser, speech-like variation of rhythmic values but often a metrical fluidity as well, the latter frequently expressed in changes of metre; rigid rhythm influenced by the *giusto* style of folk performance (to quote Bartók, 'What most interested me in the rigid rhythm . . . were the changes of measure. I fully exploited these possibilities . . . later with, perhaps, some exaggeration' (see Vinton, ibid, and Suchoff, ibid, 386); in this connection Bartók cited the first movement of the Music for Strings, Percussion and Celesta); and dotted rhythm, not only in the common ratios of

1:3 and 3:1 but more frequently in the gentler 1:2 and 2 :1. Just as Bartók's music is always tonal, even if in an enlarged sense, so it always has an essential pulse, and the barring is always the necessary outcome of melodic, harmonic, rhythmic and interpretative considerations (though an isolated irregular bar is to be seen not as an element of metrical structure but as a momentary variation, usually the temporal projection of an accent). The periodic alternation of irregular metres, or of regular with irregular, came about at a time of particular rhythmic inventiveness in Bartók's music, probably, as Breuer has suggested, under the influence of the metric and rhythmic *giusto* types in the Romanian *colinda*; it is to be seen, for example, in the First Violin Sonata, the Dance Suite and the first movement of the First Piano Concerto. The last great external influence was the so-called 'Bulgarian' rhythm; Brăiloiu has called it 'aksak' rhythm. Bartók probably turned his attention to such rhythms quite intensively at the beginning of the 1930s, and he embarked on a further revision of his Romanian collection (and others) in order to record the 'Bulgarian' micro-fluctuations. He used 'Bulgarian' rhythm compositionally in some movements that betray the influence in their titles, such as the middle movement, 'Alla bulgarese', of the Fifth Quartet. The principle here is to create units built up on groups of 2, 3, sometimes 4, in which the group of 3 is an autonomous subsidiary accent within the bar. In the 9/8 opening movement of the Sonata for two pianos and percussion, the simultaneous and successive rhythmic structures of $3 + 3 + 3$, $2 + 2 + 2 + 3$, $3 + 2 + 2 + 2$, $4 + 2 + 3$, $3 + 2 + 4$ etc give the music its dynamism. Later Bartók made use of 'Bulgarian' rhythm to enhance or make 'distant' certain themes that were trite or pseudo-quotations; this happens in the third movement of *Contrasts*, in a Hungarian-flavoured tune in 8/8 + 5/8, and also, with longer rhythmic values, in the quotation of a popular Hungarian song in the fourth movement of the Concerto for Orchestra, where the metre is 3/4 + 5/8.

But even before 'Bulgarian' rhythm had influenced such metrical devices, Bartók had endeavoured, like Stravinsky, to generate polyrhythmic, cyclical music with sections of different and often asymmetrical lengths in the different layers. At a simple level these poly-rhythms or polymetres result from the synchronization of an ostinato, or parallel ostinatos, with material consisting of developed motifs, as in the introduction to *The Miraculous Mandarin*. More complex examples can be found in moments of polymetric tension when parts of a chamber work are treated differently, being very divergently articulated and emphasized (see Mihály's analysis of the Fourth Quartet and Somfai's analytical notes on the 1926 'piano year' style in *Tizennyolc Bartók-tanulmány*, 1981, pp.164ff).

One of Bartók's rhythmic–metric principles with origins more in art music than in folk music – and exemplifying his need for proportion – is the dialectical counterpoise of the upbeat material with a characteristic Hungarian stress; the *Allegro barbaro* offers a perfect example of this. But a more fruitful idea, this time linked with the practice of proportion in early music, was the reintroduction of 'square' metres from the exposition in the reprise, or in the corresponding section of a palindromic form, but now in a dance-like, rolling triple time. In the rondo of the Second Piano Concerto, for example, the first-movement material reappears in triplets between blocks of 2/4; in the second movement of the Music for Strings, Percussion and Celesta the bulk of the 2/4 exposition returns in 3/8; and in the Violin Concerto (no.2) the 4/4 material of the first movement comes back in 3/4 in the third, bringing a specifically waltz-like lilt. Bartók also tried the reverse process, as in the Sonata for two pianos and percussion, where the 2/4 'country dance' of the third movement balances the 9/8 of the opening movement, and in the dancing 2/4 return of first-movement material (originally in 9/8 and 6/8) in the third movement of the Divertimento.

Bartókian rhythm, and the concomitant question of a correct tempo, is a matter of controversy in performance; despite the legendary precision of his scores they contain elements that depend for interpretation on the knowledge of a convention. Some of these, particularly in the piano music (the performance of rubato, the measurement of stresses and pauses etc) can probably be understood on the basis of national tradition and temperament. But Bartók was the guardian not only of the Hungarian tradition but of several others, and one performer is rarely the inheritor of all. Fortunately Bartók's own records provide a great deal of information. They show, for instance, that where a strictly uniform 'Bulgarian' rhythm is written, he made a distinction betwen true strictness and a more flexible tempo, as in his *Mikrokosmos* recordings. Difficulty in matters of tempo arises from Bartók's dual system of metronome marks and movement durations, used from the Fifth Quartet onwards, for the two sometimes do not correspond. It is known that he made numerous revisions of tempo (and other alterations) apart from those needed to correct the serious errors in the Universal publications of 1918–20. Furthermore, his recordings frequently introduce final corrections or concert alternatives. (For a study of the source value of Bartók's recordings see Somfai's article in the centenary edition of Bartók's recordings.)

CHAPTER TEN

Last works, 1938–45

Although the period 1940–42, Bartók's first American years, did not produce any significant original works, so that a natural break in his stylistic development might seem to present itself, the transformation had begun earlier: the striking individuality of his music and its general manner were altered after the completion of the Music for Strings, Percussion and Celesta and the Sonata for two pianos and percussion. It was not an abrupt change but it was irreversible, and from this point his music became less rigorous, less strictly organized, more fluid, perhaps more colourful, but definitely more confessional, more marked by extremes. A period of constructivism was followed by one of, so to speak, human idealism. Bartók continued his efforts to emulate Classical forms and genres, but the whole effect is of Romanticism.

The fact that Bartók's later works were written to commission played only a small part in this development, for, as he wrote to his elder son in 1939, 'since 1934, virtually everything I have done has been commissioned'. With the exception of *Contrasts*, where a certain jazz colouring in Bartókian style may be derived from records of the Benny Goodman band sent by Szigeti to the composer, the commission limited only the choice of medium and the technical difficulty of a work. Thus the Divertimento was required to be simple and for small forces; the Concerto for Orchestra was designed to display Koussevitzky's Boston SO; and the easier, less personal piano style of the Third Concerto was natural in a piece written by Bartók for his wife and not, like the other works for piano and orchestra, for himself.

Personal circumstances, however, both political and private, did contribute decisively to the change in style. Countless letters show that the rapid advance of Nazism plunged Bartók into a state of panic, protest and withdrawal, so that the frantic urge to complete his work alternated unpredictably with creative paralysis. Also, his mother's serious illness in summer 1939 and her death in December affected him deeply, and the coincidence of general and family griefs both determined his departure from Hungary and drove him towards that profound pessimism expressed so powerfully in the Sixth Quartet. This lament, punctuated by 'scenes from life' yet unfolding with the relentlessness of a Greek tragedy, was a work that Bartók could have produced only in an abyss of emotional upheaval or collapse, and it is significant that he was unable to realize an intended rustic dance finale (see Suchoff, 1967–8). The *mesto* conclusion to the quartet is quite exceptional: every other late multi-movement work arrives at a fervently optimistic ending, and this seems almost to have been an article of aesthetic faith with Bartók. The joyous finale might well be taken as an expression of philosophical ideals, for it is not a conventional heroic conclusion but a derivative of the dance, drawing its optimism from the common expression of people. In chamber and instrumental music it is usually of an abstract type, as in the Sonata for solo violin; in orchestral works it is folklike,

rustic and ironically humorous, with varied contrapuntal episodes or passages of pathos, as in the 1937–8
Violin Concerto, the Third Piano Concerto and the
draft Viola Concerto. The Concerto for Orchestra ends
with a characteristically 'multi-national' round-dance.

Though Bartók's view of a 'brotherhood of nations'
was expressed thus in formal terms, it is an important
feature of his late style that an opening theme is more
often a grand instrumental melody of frankly Hungarian
tone. The best example is in the opening movement of
the Violin Concerto, headed 'Tempo di verbunkos' in an
early manuscript; other examples are the 'Verbunkos'
first movement of *Contrasts* and the first movement of
the Third Piano Concerto. Avowedly Hungarian themes
and rhythms are also more prominent in the central
slow monologues of, for example, the Divertimento
(second movement, fugato in dotted rhythm and the
apotheosis beginning in bar 64) and the Concerto for
Orchestra ('Elegia', two laments from bars 34 and 62).
Other movements show that Bartók's forms could still
suggest in extreme cases a background programme, for
example the fourth movement of the Concerto for
Orchestra (see Ujfalussy, 1965, for various interpretations) and the second of the Third Piano Concerto, where
the reference is more identifiable, in a quotation from
the 'Heiliger Dankgesang' of Beethoven's op.132. (This
explains Bartók's marking 'Adagio religioso', which
Serly used again in his edition of the Viola Concerto.) In
this late period Bartók made more use of orthodox
forms (e.g. in the finely proportioned sonata structures
of the Sixth Quartet and the Third Piano Concerto) and
of genres and textures from Baroque music (e.g. in the
Bachian conception of the Sonata for solo violin and in

the concerto grosso aspects of the Divertimento and the Violin Concerto).

The brilliant orchestral colouring of Bartók's late works, something new in his music, undoubtedly played a part in bringing them great posthumous public favour. After the enthrallingly inventive scoring of *The Miraculous Mandarin* Bartók had, following Stravinsky, tried a new conception in the First and Second Piano Concertos: the latter's opening movement uses an orchestra of wind and percussion, the slow movements have a fuller but smaller complement, and the whole ensemble is used only in the finales. But he did not follow this up. For a while he made new departures in scoring for strings, percussion and piano, separately or combined. Then in the Violin Concerto he introduced a new style of instrumentation, particularly in the middle movement where the melodic material is scored with maturity and delicacy and there are important wind solos (see Somfai, *SM*, 1977). These tendencies were reinforced in the American scores. It was not, of course, a progressive orchestral style for its time, but it was at least as individual and enduring as the more constructive colouring of the classical middle period. And it was symptomatic of Bartók's conscious desire to make contact with his audience in these last compositions. He was not above providing alternative endings from which the performers might choose, in the Violin Concerto and the Concerto for Orchestra. Such 'concessions' have been considered by some commentators to have been stimulated by American taste or by Bartók's disappointment at the neglect of his works, but, as has been noted, the late developments preceded his emigration and the American works are too few to show any further evolution.

During the almost silent period of 1940–42 Bartók turned down a number of interesting requests for compositions from his new publisher, Ralph Hawkes, on the grounds of creative impotence. The works he subsequently wrote are each somewhat exceptional. If only because of its genre, the Concerto for Orchestra stands by itself; also, it was made up, quite unusually, of elements composed earlier than the date on the score and intended for a ballet. The Sonata for solo violin contains the quintessence of Bartókian melody and strives for the utmost concentration of expression, so that it cannot be considered typical. The Third Piano Concerto, composed for Bartók's wife and cast in a clear Mozartian mould, is unique in Bartók's output. However, there are signs of a creative upsurge in the ambitious plans that Bartók had at the time of his death. According to his own account the Viola Concerto was 'ready in draft' and its orchestration would have been 'a purely mechanical feat' for him; he was preparing sketches for a seventh quartet; and he wanted to write a new choral work and orchestrate the Romanian Folk Dances (see Kroó, 1970). So one can understand his statement to a hospital doctor: 'I am only sorry that I have to leave with my baggage full'.

Bartók's life as an artist and his influence were quite different from those of Schoenberg and Stravinsky, his two great contemporaries. He was not a brilliant, autocratic teacher and theorist; nor did he enter the international arena in the important decades between 1910 and 1930. At the time he was not considered unreservedly a leading figure in Western music. His few striking contemporary successes and his high posthumous popularity did not and cannot disguise the fact that his influence was relatively minor; he left no school, only a few epigones. His greatness lies not so much in his technical and stylistic innovations as in his extraordinary aptitude for creative synthesis – and, finally, in the ideal he formed and realized that modern, alienated man might create deeply personal art by turning to some pure and precious source.

Catalogues: D. Dille: *Thematisches Verzeichnis der Jugendwerke Béla Bartóks 1890–1904* (Budapest, 1974) [DD]

A. Szöllösy: 'Bibliographie des oeuvres musicales et écrits musicologiques de Béla Bartók', in B. Szabolcsi, ed.: *Bartók: sa vie et son oeuvre* (Budapest, 1956) [rev. nos. from Eng. trans, 1971, of J. Ujfalussy (1965)] [Sz]

Publishers: Boosey & Hawkes [B]
Dover [Do]
Magyar Kórus [M]
Rozsnyai Károly [R]
Rózsavölgyi es Társa [Rv]
Universal [U]
Zeneműkiadó Vállalat (Editio Musica) [Z]

Numbers in the right-hand column denote references in the text.

STAGE

Sz	op.	Title	Genre (acts, libretto/scenario)	Composition	First performance	Publication, Remarks	
48	11	A Kékszakállú herceg vára [(Duke) Bluebeard's castle]	opera (1, B. Balázs)	1911, rev. 1912, 1918	cond. E. Tango, Budapest, Opera, 24 May 1918	vocal score, U 1922; full score, U 1925, 1963	14, 15, 18, 23, 25, 39, 44, 50, 51, 52, 53
60	13	A fából faragott királyfi [The wooden prince]	ballet (1, Balázs)	1914–16, orchd 1916–17	cond. Tango, Budapest, Opera, 12 May 1917	pf score, U 1921; full score, U 1924; suite, Sz 60; suite, 1932, unpubd; see also 'Orchestral'	18, 22–3, 25, 44, 50, 51, 52
73	19	A csodálatos mandarin [The miraculous mandarin]	pantomime (1, M. Lengyel)	1918–19; orchd 1923, rev. 1924, 1926–31	cond. E. Szenkár, Cologne, Stadttheater, 27 Nov 1926	pf 4 hands score, U 1925; full score, U 1955; suite, Sz 73; see also 'Orchestral'	20, 23, 25, 51, 54, 55, 70, 76

ORCHESTRAL

Sz	op.	Title, Scoring	Composition	First performance	Publication	Remarks	
—	—	Valcer	c1900			DD 60b; arr. of pf work, DD 60a/1–2	
—	—	Scherzo, Bb	c1901			DD 65	
—	—	Symphony, Eb	1902, orchd 1903	scherzo, cond. I. Kerner, Budapest, 29 Feb 1904		DD 68; scherzo, C, orchd; other movts in sketch	8, 34
—	—	Kossuth, sym. poem	1903	Budapest, cond. Kerner, 13 Jan 1904	Z 1963	DD 75a; tableau 10 arr. pf; DD 75b	8, 9, 35

Sz	op.	Title. Scoring	Composition	First performance	Publication	Remarks	
27	1	Rhapsody, 2nd version, pf, orch	?1904	Bartók, cond. C. Chevillard, Paris, ?3–8 Aug 1905	Rv 1910, Z 1954	arr. from pf piece, Sz 26; arr. 2 pf, 1905 (Rv 1910, 1919; Z 1955)	9
28	2	Scherzo (Burlesque), pf, orch	1904	E. Tusa, cond. G. Lehel, Budapest, 28 Sept 1961	Z 1961	arr. 2 pf, unpubd	9, 36, 52
31	3	Suite no.1, full orch	1905, rev. c1920	movts 1, 3–5, Vienna, 29 Nov 1905; complete, cond. Hubay, Budapest, 1 March 1909	Rv 1912, rev. Z 1956		9, 11, 35–6
34	4	Suite no.2, small orch	movts 1–3, 1905, movt 4, 1907; rev. 1920, 1943	movt 2, Berlin, 2 Jan 1909; complete, cond. Kerner, Budapest, 22 Nov 1909	Bartók 1907, U 1921, B 1948	arr. 2 pf, Sz 115a	11, 29, 35, 36
36	—	Violin Concerto ('no.1')	1907–8	H. Schneeberger, cond. P. Sacher, Basle, 30 May 1958	B 1959	1st movt rev. as no.1 of Two Portraits; 2nd movt arr. vn, pf, unpubd	12, 14, 36, 49
37	5	Két portré [Two portraits] 1 Egy ideális [One ideal] 2 Egy torz [One grotesque]	no.1, 1907–8; no.2, orchd 1911	no.1, Budapest, 12 Feb 1911; complete, cond. I. Strasser, Budapest, 20 April 1916	R 1911 or 1912, B 1950, Z 1953	no.1 from 1st movt Vn Conc., Sz 36; no.2, arr. of pf work, Sz 38/14	14, 39, 47, 49
46	10	Két kép [Two pictures] 1 Virágzás [In full flower] 2 A falu tánca [Village dance]	1910	cond. Kerner, Budapest, 25 Feb 1913	Rv 1912, B 1950, Z 1953	arr. pf, c1911 (Rv 1912, B 1950, Z 1953, Do 1981)	12, 42, 49
47a	—	Román tánc [Romanian dance]	1911	cond. L. Kun, Budapest, 12 Feb 1911	Z 1965	arr. of pf work, Sz 43/1	
51	12	Four Pieces 1 Preludio, 2 Scherzo, 3 Intermezzo, 4 Marcia funebre	1912, orchd 1921	cond. Dohnányi, Budapest, 9 Jan 1922	U 1923		15–17, 22, 51–2, 55
60	13	The Wooden Prince, suite	?1921–4	cond. Dohnányi, Budapest, 23 Nov 1931		3 dances from ballet	
68	—	Román népi táncok [Romanian folkdances]	1917	cond. E. Lichtenberg, Budapest, 11 Feb 1918	U 1922	arr. of pf work, Sz 56	

73	—	The ... Mandarin, suite	1919, 192?	cond. Dohnányi, Budapest, 15 Oct 1928	U 192?		
77	—	Táncszvit [Dance suite], orig. title: Tánc-Suite	1923	cond. Dohnányi, Budapest, 19 Nov 1923	U 1924	1 omitted no. in draft; arr. pf, 1925 (U 1925)	22, 23, 45, 51, 55, 56, 59, 69
83	—	Piano Concerto no.1	1926	Bartók, cond. Furtwängler, Frankfurt am Main, 1 July 1927	U 1927, 1928	arr. 2 pf (U 1927)	23, 24, 59, 60, 69, 76
87	—	Rhapsody no.1, vn, orch	1928	Szigeti, cond. Scherchen, Königsberg, 1 Nov 1929	U 1929	arr. of vn, pf work, Sz 86	23
90	—	Rhapsody no.2, vn, orch	1928, rev. 1944	Székely, cond. Dohnányi, Budapest, 26 Nov 1929	U 1929, B 1949	arr. of vn, pf work, Sz 89	23
95	—	Piano Concerto no.2	1930–31	Bartók, cond. Rosbaud, Frankfurt am Main, 23 Jan 1933	U 1932, 1955	arr. 2 pf (U 1941)	23, 25, 27, 60, 61, 62, 63, 70–71, 76
96	—	Erdélyi táncok [Transylvanian dances]	1931	cond. M. Freccia, Budapest, 24 Jan 1932	Rv 1932, Z 1955	arr. of Pf Sonatina, Sz 55	
97	—	Magyar képek [Hungarian sketches]	1931	nos.1–3, 5, Budapest, 24 Jan 1932; ?1st complete, cond. H. Laber, Budapest, 26 Nov 1934	R-Rv 1932, Z 1954	arr. of pf works, Sz 39/5, 39/10, 45/2, 47/2, 42/ii/42	64–5
100	—	Magyar parasztdalok [Hungarian peasant songs]	1933	cond. G. Baranyi, Szombathely, 18 March 1934	U 1933	arr. of pf works, Sz 71/6–12, 14–15	
106	—	Music for Strings, Percussion and Celesta	1936	cond. Sacher, Basle, 21 Jan 1937	U 1937		26, 44, 45, 62, 66–9, 71, 73
112	—	Violin Concerto ('no.2')	1937–8	Székely, cond. Mengelberg, Amsterdam, 23 March 1939	B 1946	arr. vn, pf (B 1941)	26, 55, 63, 64, 71, 75, 76
113	—	Divertimento, str	1939	cond. Sacher, Basle, 11 June 1940	B 1940		26, 44, 56, 63, 71, 73, 75, 76

Sz	op.	Title. Scoring	Composition	First performance	Publication	Remarks	
115	—	Two Piano Concerto	1940	Kentner, Kabos, cond. Boult, London, 14 Nov 1942	B 1970	arr. of Sonata, 2 pf, perc, Sz 110	29
116	—	Concerto for Orchestra	[?1942–] 1943, rev. Feb 1945	cond. Koussevitzky, Boston, 1 Dec 1944	B 1946	arr. pf for planned ballet, Jan 1944, unpubd	30, 56, 70, 73, 75, 76, 77
119	—	Piano Concerto no.3	1945	G. Sándor, cond. Ormandy, Philadelphia, 8 Feb 1946	B 1947	last 17 bars scored by Serly	31, 73, 75, 77
120	—	Viola Concerto	1945	Primrose, cond. Dorati, Minneapolis, 2 Dec 1949	B 1950	completed from draft by Serly; vc version (B 1956)	31, 75, 77

VOCAL ORCHESTRAL

Sz	Title	Text	Scoring	Composition	First performance	Publication, Remarks	
—	Tiefblaue Veilchen	C. Schoenaich-Carolath	S, orch	1899		DD 57	
79	Falun (Tri dedinské scény) [Three village scenes]	Slovakian trad.	4/8 female vv, chamber orch	1926	cond. Koussevitzky, New York, 1 Feb 1927	full score, U 1927; vocal score, U 1927; arr. of songs, Sz 78/3–5	59
94	Cantata profana (A kilenc csodaszarvas) [The nine enchanted stags]	Rom. colinda, arr. and Hung. trans. Bartók	T, Bar, double chorus, orch	1930	cond. A. Buesst, London, 25 May 1934	full score, U 1934, 1957; vocal score, U 1934, 1951	24, 39, 44, 58, 62, 65, 66
101	Magyar népdalok [Five Hungarian folksongs]	Hung. trad.	1v, orch	1933	cond. Dohnányi, Budapest, 23 Oct 1933	arr. of songs, Sz 92/1, 2, 11, 14, 12	

— Est [Evening], DD 74 (K. Harsányi), 8 male vv, 1903 (Z 1965)

50 Négy régi magyar népdal [4 old Hungarian folksongs], 4 male vv, 1910, rev. 1912 (U 1928): Rég megmondtam bús gerlice [Long ago I told you]; Jaj Istenem, kire várok [O God, why am I waiting?]; Angyomasszony kertje [In my sister-in-law's garden]; Béreslegény [Farmboy, load the cart well]

58 Two Romanian Folksongs, 4 female vv, ?1915, completed from draft by B. Suchoff. Nu te supăra mireasă [On her wedding day]; Măi badijă prostule [Fickle lover, silly man]   17

69 Tót népdalok [Slovácké ľudové piesne] [Slovak folksongs], 4 male vv, 1917 (U 1918): Ej, posluchajte málo [Ah, listen now my comrades]; Ked'ja smutny pojdem [Back to fight]; Kamarádi moji [War is in our land]; Ej, a ked'mna zabiju [Ah, if I fall in battle]; Ked'som šiou na vojnu [Time went on]

70 Négy tót népdal [Štyri slovenské piesne] [4 Slovak folksongs], 4vv, pf, 1917 (U 1927, Z 1950): Zadala mamka [Wedding song]; Naholi, naholi [Song of the hay-harvesters]; Rada pila, rada jedla [Song from Medzibrod]; Gajdujte, gajdence [Dancing song]

93 Magyar népdalok [Hungarian folksongs], mixed vv, 1930 (U 1932): Elhervadt cidrusfa [The prisoner]; Ideje bujdosásimnak [The wanderer]; Adj el, anyám [Finding a husband]; Sarjut eszik az ökröm [My ox is grazing]; Az én lovam szajkó [Love-song]   24

99 Székely dalok [Székely songs], 6 male vv, 1932 (nos.3–5 Schweizerische Sängerzeitung (1933), nos.1–2 M 1938, complete Z 1955, rev. 1982): Hej, de sokszor megbántottál, Túl vagy rózsám [How often I've grieved for you, It's all over]: Istenem, életem nem igen gyönyörű [My God, my life]; Vékony cérna, kemény mag [Slender thread, hard seed]; Kilyénfalvi középtizbe [Girls are gathering in Kilyénfalva]; Vékony cérna, kemény mag; Járjad pap a táncot [Do a dance, priest]   24

103 27 Two- and Three-part Choruses (Hung. trad., arr. Bartók), children's vv (vols.i–vi), female vv (vols.vii–viii), 1935–6 (M 1937, 1941; Z 1953, 1983; 9 nos. B 1955; other 18 nos. Z 1972)   26, 64, 66
vol.i: Tavasz [Spring]: Ne hagyj itt! [Only tell me]; Jószág-igéző [Enchanting song]

vol.ii: Leventék az otthonhoz [To my homeland]; Játék [Game] [Choosing of a girl]; Héjja, héjja, karahéjja [Thieving bird]
vol.iii: Ne menj el [Don't leave me]; Van egy gyürüm [The fickle girl]; Senkim a világon [Song of loneliness]; Cipósütés [Bread-baking]
vol.iv: Huszármóta [Hussar]; Resteknek nótája [Loafer]; Bolyongás [Lonely wanderer]; Lánycsúfoló [Mocking of girls]
vol.v: Legénycsúfoló [Mocking of youth]; Mihálynapi köszöntő [Michaelmas greeting]; Leánykérő [The wooing of a girl]
vol.vi: Keserves [Lament]; Madárdal [Song of the bird]; Csujogató [Stamping feet]
vol.vii: Bánat [The sorrow of love]; Ne láttalak volna! [Had I never seen you]; Elment a madárka [The song-bird's promise]
vol.viii: Párnás táncdal [Pillow dance]; Kánon; Isten veled! [Lover's farewell]   66
nos.iv/1, iii/1, iv/2, iv/3, iii/4 arr. with school orch (M 1937; Z 1962, 1963); nos.i/2, v/1 arr. with small orch (B 1942)

104 Elmult időkből [From olden times] (Hung. trad., arr. Bartók), 3 male vv, 1935 (M 1937): Nincs boldogtalanabb [No-one's more unhappy than the peasant]; Egy, kettő, három, négy [One, two, three, four]; Nincsen szerencsésebb [No-one is happier than the peasant]

CHAMBER

— A Duna folyása [The course of the Danube], DD 20b, vn, pf [arr. of pf work, DD 20a], 1894; pf part lost
— Sonata, c, op.5, DD 37, vn, pf, 1895   33
— Violin pieces, op.7, DD 39, 1895, lost; 2 fantasias, opp.8–9, DD 40–41, 1896, lost
— String Quartet no.1, B, op.10, DD 42, 1896, lost   33
— String Quartet no.2, c, op.11, DD 43, 1896, lost   33
— Piano Quintet, C, op.14, DD 46, 1897, lost   33
Sonata, A, op.17, DD 49, vn, pf, 1897; pf part of 2nd movt only in sketch   33
— Piano Quartet, c, op.20, DD 52, 1898
— String Quartet, F, DD 56, 1898   3, 33
— Piano Quintet, DD B10 and B12, frags., ?1899   33

Scherzo in sonata form, DD 58, str qt, 1899–1900

Duo (Canon), DD 69, 2 vn, 1902

Andante (Albumblatt), A, DD 70, vn, pf, 1902 (Z 1980); MS pubd, ed. L. Somfai (Budapest, 1980) — 7, 7

Andante, F♯, DD B14, vn, pf, ?1903; inc.

Sonata, e, DD 72, vn, pf, 1903 (Documenta Bartókiana, 1964–5; Z 1968) — 8, 9, 34

Piano Quintet, DD 77, 1903–4, rev. ?1920 (Z 1970) — 8, 9, 15, 35

35 Gyergyóból [From Gyergyó], rec, pf [arr. for pf, Sz 35a], 1907 (Z 1961)

40 String Quartet no.1, op.7, 1908 (Rv 1910; Z 1956, 1964) — 12, 15, 21, 39, 47, 48

67 String Quartet no.2, op.17, 1915–17 (U 1920) — 18, 19, 45, 51, 52

75 Sonata no.1 [MS: op.21], vn, pf, 1921 (U 1923) — 21, 55–6, 58, 69

76 Sonata no.2, vn, pf, 1922 (U 1923) — 21, 44, 55, 56, 58

85 String Quartet no.3, 1927 (U 1929) — 23, 58, 61

86 Rhapsody no.1, vn, pf, 1928 (U 1929); orchd, Sz 87; arr. vc, pf, Sz 88 — 23, 65

88 Rhapsody, vc, pf, 1928 (U 1930) — 23

89 Rhapsody no.2, vn, pf, 1928 (U 1929), rev. 1945 (B 1947); orchd, Sz 90 — 23, 65

91 String Quartet no.4, 1928 (U 1929) — 23, 45, 50, 60, 61, 62, 63, 65, 70

98 44 Duos, 2 vn, 1931 (7 nos. Schott 1932, complete U 1933) — 23, 64
vol.i: 1 Párosító [Teasing song]; 2 Kalamajkó [Dance]; 3 Menuetto; 4 Szentiványéji [Midsummer night song]; 5 Tót nóta [Slovak song]; 6 Magyar nóta [Hungarian song]; 7 Oláh nóta [Romanian song]; 8 Tót nóta [Tót song]; 9 Játék [Play]; 10 Rutén nóta [Ruthenian song]; 11 Gyermekrengetéskor [Lullaby]; 12 Szénagyűjtéskor [Hay-harvesting song]; 13 Lakodalmas [Wedding song]; 14 Párnás-tánc [Cushion dance]
vol.ii: 15 Katonanóta [Soldier's song]; 16 Burleszk; 17 Menetelő nóta [Marching song]; 18 Menetelő nóta; 19 Mese [Fairy tale]; 20 Dal [Song]; 21 Újévköszöntő [New year's greeting]; 22 Szúnyogtánc [Mosquito dance]; 23 Menyasszony-búcsúztató [Wedding song]; 24 Tréfás nóta [Gay song]; 25 Magyar nóta [Hungarian song]; 26 Ugyan édes komámasszony [Teasing song]; 27 Sántatánc [Limping dance]; 28 Bánkódás [Sorrow]; 29 Újév-

köszöntő; 30 Újévköszöntő; 31 Újévköszöntő; 32 Máramarosi tánc [Dance from Máramaros]; 33 Aratáskor [Harvest song]; 34 Számláló nóta [Counting song]; 35 Rutén kolomejka [Ruthenian kolomejka]; 36 Szól a duda [Bagpipes]
vol.iv: 37 Preludium és kánon; 38 Forgatós (Invátita bátránilor) [Romanian whirling dance]; 39 Szerb tánc (Zaplet) [Serbian dance]; 40 Oláh tánc [Wallachian dance]; 41 Scherzo; 42 Arab dal; 43 Pizzicato; 44 Erdélyi tánc (Ardeleana) [Transylvanian dance] — 8, 9, 34

nos.28, 38, 43, 16, 36, 32 arr. pf, Sz 105 — 26, 61, 62, 63, 69, 71

102 String Quartet no.5, 1934 (U 1936) — 26, 68, 69, 71, 73

110 Sonata, 2 pf, 2 perc, 1937 (B 1942) — 26, 28, 44, 70, 73,

111 Contrasts, vn, cl, pf, 1938 (B 1942) — 75

114 String Quartet no.6, 1939 (B 1941) — 28, 51, 55, 74, 75

117 Sonata. vn, 1944 (B 1947); ¼-tone version of last movt. unpubd — 30, 55, 74, 75, 77

(authorized arrangements)

Magyar népdalok [Hungarian folksongs], vn, pf [arr. of pf work, Sz 42: vols.i–ii: nos.34, 36, 17, 31, 16, 14, 19, 8, 21], 1934, arr. T. Országh, Bartók (R 1934, Z 1954)

Ungarische Volksweisen, vn, pf [arr. of pf work, Sz 42: vols.i–ii: nos.28, 18, 42, 33, 6, 13, 38], arr. Szigeti (U 1927)

Sonatina, vn, pf [arr. of pf work, Sz 55], arr. E. Gertler (Rv 1931)

Romanian Folkdances, vn, pf [arr. of pf work, Sz 56], arr. Székely (U 1926)

PIANO — 32

Walczer, op.1, DD 1, 1890; Változó darab [Variable piece], op.2, DD 2, 1890; Mazurka, op.3, DD 3, 1890; A budapesti tornaverseny [Gymnastic contest in Budapest], op.4, DD 4, 1890; Sonatina no.1, op.5, DD 5, 1890; Oláh darab [Wallachian piece], op.6, DD 6, 1890; Gyorspolka [Fast polka], op.7, DD 7, 1891; 'Béla' polka, op.8, DD 8, 1891; 'Katinka' polka, op.9, DD 9, 1891; Tavaszi hangok [Sounds of spring], op.10, DD 10, 1891; 'Jolán' polka, op.11, DD 11, 1891; 'Gabi' polka, op.12, DD 12, 1891; Nefelejts [Forget-me-not], op.13, DD 13, 1891; Ländler no.1, op.14, DD 14, 1891; 'Irma' polka, op.15, DD 15, 1891; Radegundi visszhang [Echo of Radegund], op.16, DD 16, 1891; Induló [March], op.17, DD 17, 1891; Ländler no.2, op.18, DD 18, 1891; Cirkusz polka, op.19, DD 19, 1891; A Duna folyása [The course of — 1, 1, 2, 2, 32 *the Danube]*, op.20, DD 20a, 1890–94, arr. vn, pf; Sonatina no.2, op.21, DD 21, ?1891; Ländler no.3, op.22, DD 22, 1892, lost; Tavaszi dal [Song of spring], op.23, DD 23, 1892; 'Margit' polka, op.25, DD 25, 1893; 'Ilona' mazurka, op.26, DD 26, 1893; 'Loli' mazurka, op.27, DD 27, 1893; 'Lajos' valczer, op.28, DD 28, 1893; 'Elza' polka, op.29, DD 29, 1894; Andante con variazioni, op.30, DD 30, 1894: X.Y., op.31, DD 31, 1894, lost; Sonata no.1, g, op.1, DD 32, 1894; Scherzo, g, DD 33, 1894; Fantasie, a, op.2, DD 34, 1895; Sonata no.2, F, op.3, DD 35, 1895; Capriccio, b, op.4, DD 36, 1895; Sonata no.3, C, op.6, DD 38, 1895, lost; Andante, Scherzo and Finale, op.12, DD 44, 1897, lost — *33, 33*

— Drei Klavierstücke, b, C, ab, op.13, DD 45, 1897 (no.1 Z 1965) — *3*

— Two Pieces, op.15, DD 47, 1897, lost

— Great Fantasy, op.16, DD 48, 1897, lost

— Scherzo (Fantasie), B, op.18, DD 50, 1897 (Z 1965)

— Sonata, op.19, DD 51, 1898, lost

— Drei Klavierstücke, op.21, DD 53 (nos.1–2 Z 1965)

— Scherzo, b, DD 55, 1898;

— Scherzo, bb, DD 59, c1900

— Six Dances, DD 60a, c1900 (no.1, facs., 1913); nos.1–2 orchd, DD 60b — *7*

— Scherzo, bb, DD 63, 1900

— Változatok [12 variations], DD 64, 1900–01 (Z 1965) — *7*

— Tempo di minuetto, DD 66, 1901

— Four Pieces, DD 71, 1903 (Bárd 1904; nos.1–3 B 1950; Z 1956, 1965, Do 1981): Study for the Left Hand; Fantasy I; Fantasy II; Scherzo — *6, 8, 9, 35*

— Marche funèbre, DD 75b [arr. of Kossuth: tableau 10], 1903 (Magyar Lant 1905, R ?1910, Z 1953, Do 1981) — *26*

26 Rhapsody, op.1, 1904 (Adagio mesto Rv 1908; complete Rv 1923, Z 1955, Do 1981); arr. pf orch, Sz 27 — *6, 21, 24, 35*

— Petits morceaux [free arr. of songs, Sz 29/2, DD 67/1], ?1905–7 (Z 1965)

35a Három Csik megyei népdal [3 Hungarian folksongs from the Csik district], 1907 (R 1910, B 1950, Z 1954, Do 1981) — *12, 42*

38 14 Bagatelles, op.6, 1908 (R 1908, B 1950, Z 1953, Do 1981): Molto sostenuto; Allegro giocoso; Andante; Grave [arr. of folksong Mikor gulyásbojtár voltam]; Vivo [arr. of folksong Ej! po pred naš, po pred naš]; Lento; Allegretto molto capriccioso; Andante sostenuto; Allegretto grazioso; Allegro; Allegretto molto rubato; Rubato; Elle est morte (Lento — *12, 14, 15, 21, 40, 41, 42, 43, 47, 48*

funebre); Valse: ma mie qui danse (Presto); no.14 orchd, Sz 37/2 — *12, 14, 47*

39 Ten Easy Pieces, 1908 (R 1908, B 1950, Z 1951, Do 1981), with Ajánlás [Dedication]; Paraszti nóta [Peasant song]; Lassú vergődés [Frustration]; Tót legények tánca [Slovakian boy's dance]; Sostenuto; Este a székelyeknél [Evening in Transylvania [Evening with the Széklers]]; Gödöllei piacčeren leesett a hó [Hungarian folksong]; Hajnal [Dawn]; Azt mondják, nem adnak [Slovakian folksong]; Ujjgyakorlat [Five-finger exercise]; Medvetánc [Bear dance]; nos.5, 10 orchd, Sz 97/1, 2 — *12, 14, 39, 40*

41 Két elégia [2 elegies], op.8b, 1908, 1909 (R 1910, B 1950, Z 1955, Do 1981) — *12*

42 Gyermekeknek [For children], 85 pieces, i–iv [i–ii after Hung., iii–iv after Slovakian folksongs], 1908–9 (R 1910–12, Z 1950, Do 1981), rev. 1945, 79 pieces, i–ii (B 1947); ii/42 orchd, Sz 97/5; i/16 arr. 1v, pf, Sz 109 — *42*

43 Két román tánc [2 Romanian dances], op.8a, 1909–10 (Rv 1910, B 1950, Z 1951, Do 1981); MS pubd, ed. L. Somfai (Budapest, 1974); no.1 orchd, Sz 47a — *12, 14, 39, 42, 44*

44 Vázlatok [7 sketches], op.9b, 1908–10 (R 1912, Augener 1950, B 1950, Z 1954, Do 1981): Leányi arckép [Portrait of a girl]; Hinta palinta [See-saw, dickory-daw]; Lento; Non troppo lento; Román népdal [Romanian folksong]; Oláhos [In Wallachian style]; Poco lento — *39*

45 Négy siratóének [4 dirges], op.9a, 1909–10 (Rv 1912, Delkas 1945, B 1950, Z 1955, Do 1981); no.2 orchd, Sz 97/3 — *12, 14, 39, 47*

47 Három burleszk [3 burlesques], op.8c (Rv 1912, B 1950, Z 1954, Do 1981): Perpatvar [Quarrel], 1908; Kicsit ázottan [A bit drunk], 1911; Molto vivo capriccioso, 1910; no.2 orchd, Sz 97/4 — *12, 48, 70*

49 Allegro barbaro, 1911 (U 1918) — *17*

53 Kezdők zongoramuzsikája [The first term at the piano], 18 pieces, 1913 (in pf method of Bartók and S. Reschofsky, 1913; Rv 1929, B 1950, Z 1952, Do 1981) — *17*

55 Sonatina, 1915 (Rv 1919), rev. after 1930 (B 1950, Z 1952); orchd, Sz 96 — *48, 77*

56 Román népi táncok [Romanian folkdances], 1915 (U 1918); orchd, Sz 68 — *54*

57 Román kolinda-dallamok [Romanian Christmas carols], 20 pieces in 2 sers, 1915 (U 1918)

62 Suite, op.14, 1916 (U 1918; abandoned 5th movt in Új zenei szemle, v, 1955) — *18, 20, 44–5, 51, 52*

66 Three Hungarian Folktunes, c1914–18 (no.1 in early version, Sz 65, in *Periszkóp* (1925), June–July; complete in album *Homage to Paderewski*), rev. ?1942 (B 1942): Leszállott a páva [The peacock]; Jánoshidi vásártéren [At the Jánoshida fairground]; Fehér liliomszál [White lily]

71 Tizenöt magyar parasztdal [15 Hungarian peasant songs], 1914–18 (U 1920, B 1948): 1–4 Négy régi keserves ének [4 old tunes]; 5 Scherzo; 6 Ballade (Tema con variazioni); 7–15 Régi táncdalok [Old dance tunes]; nos.6–12, 14–15 orchd, Sz 100 — 18, 54

72 [3] Studies, op.18, 1918 (U 1920) — 20, 55

74 [8] Improvisations on Hungarian Peasant Songs, op.20, 1920 (U 1922) — 21–2, 54

77 Táncszvit [Dance suite], 1925, arr. of orch suite — 59

80 Sonata, 1926 (U 1927); MS pubd, ed. L. Somfai (Budapest, 1980) — 22, 60

81 Szabadban [Out of doors], 1926 (U 1927); Sippal, dobbal [With drums and pipes]; Barcarolla; Musettes; Az éjszaka zenéje [Night music]; Hajsza [The chase] — 22, 23, 60

82 Nine Little Pieces, 1926 (U 1927): 1–4 Négy párbeszéd [4 dialogues]; 5 Menuetto; 6 Dal [Air]; 7 Marcia delle bestie; 8 Csörgő-tánc [Tambourine]; 9 Preludio – All' ungherese — 23, 59, 60

84 Három rondó népi dallamokkal [3 rondos on [Slovak] folktunes], no.1 1916, nos.2-3 1927 (U 1930) — 62

105 Petite suite [arr. of chamber work, Sz 98/28, 38, 43, 16, 36], 1936 (U 1938); 6th movt [arr. of Sz 98/32], unpubd

107 Mikrokosmos, progressive pieces, 1926, 1932–9 (B 1940) — 26, 28, 64, 71
vol.i: 1–6 Six Unison Melodies; 7 Dotted Notes; 8 Repetition; 9 Syncopation; 10 With Alternate Hands; 11 Parallel Motion; 12 Reflection; 13 Change of Position; 14 Question and Answer; 15 Village Song; 16 Parallel Motion and Change of Position; 17 Contrary Motion; 18–21 Four Unison Melodies; 22 Imitation and Counterpoint; 23 Imitation and Inversion; 24 Pastorale; 25 Imitation and Inversion; 26 Repetition; 27 Syncopation; 28 Canon at the Octave; 29 Imitation Reflected; 30 Canon at the Lower Fifth; 31 Little Dance in Canon Form; 32 In Dorian Mode; 33 Slow Dance; 34 In Phrygian Mode; 35 Chorale; 36 Free Canon; Appendix: Exercises
vol.ii: 37 In Lydian Mode; 38–39 Staccato and Legato; 40 In Yugoslav Mode; 41 Melody with Accompaniment; 42 Accompaniment in Broken Triads; 43 In Hungarian Style, 2 pf; 44 Contrary Motion, 2 pf; 45 Meditation; 46 Increasing–Diminishing; 47 Big Fair; 48 In Mixolydian Mode; 49 Crescendo–Diminuendo; 50 Minuetto; 51 Waves; 52 Unison Divided; 53 In Transylvanian Style; 54 Chromatic; 55 Triplets in Lydian Mode, 2 pf; 56 Melody in Tenths; 57 Accents; 58 In Oriental Style; 59 Major and Minor; 60 Canon with Sustained Notes; 61 Pentatonic Melody; 62 Minor Sixths in Parallel Motion; 63 Buzzing; 64 Line and Point; 65 Dialogue, 1v, pf; 66 Melody Divided; Appendix: Exercises
vol.iii: 67 Thirds against a Single Voice; 68 Hungarian Dance, 2 pf; 69 Chord Study; 70 Melody against Double Notes; 71 Thirds; 72 Dragon's Dance; 73 Sixths and Triads; 74 Hungarian Song, 1v, pf; 75 Triplets; 76 In Three Parts; 77 Little Study; 78 Five-tone Scale; 79 Hommage à J.S.B.; 80 Hommage à R. Sch.; 81 Wandering; 82 Scherzo; 83 Melody with Interruptions; 84 Merriment; 85 Broken Chords; 86 Two Major Pentachords; 87 Variations; 88 Duet for Pipes; 89 In Four Parts; 90 In Russian Style; 91–2 Two Chromatic Inventions; 93 In Four Parts; 94 Tale; 95 Song of the Fox, 1v, pf; 96 Stumblings; Appendix: Exercises
vol.iv: 97 Notturno; 98 Thumb Under; 99 Crossed Hands; 100 In the Style of a Folksong; 101 Diminished Fifth; 102 Harmonics; 103 Minor and Major; 104 Through the Keys; 105 Playsong; 106 Children's Song; 107 Melody in the Mist; 108 Wrestling; 109 From the Island of Bali; 110 Clashing Sounds; 111 Intermezzo; 112 Variations on a Folktune; 113 Bulgarian Rhythm (i); 114 Theme and Inversion; 115 Bulgarian Rhythm (ii); 116 Melody; 117 Bourrée; 118 Triplets in 9/8 Time; 119 Dance in 3/4 Time; 120 Fifth Chords; 121 Two-part Study; Appendix: Exercises
vol.v: 122 Chords Together and Opposed; 123 Staccato and Legato; 124 Staccato; 125 Boating; 126 Change of Time; 127 New Hungarian Folksong, 1v, pf; 128 Peasant Dance; 129 Alternating Thirds; 130 Village Joke; 131 Fourths; 132 Major Seconds Broken and Together; 133 Syncopation; 134 Three Studies in Double Notes; 135 Perpetuum mobile; 136 Whole-tone Scale; 137 Unison; 138 Bagpipe; 139 Merry Andrew
vol.vi: 140 Free Variations; 141 Subject and Reflection; 142 From the Diary of a Fly; 143 Divided Arpeggios; 144 Minor Seconds, Major Sevenths; 145 Chromatic Invention; 146 Ostinato; 147 March; 148–53 Six Dances in Bulgarian Rhythm

108 Seven Pieces from Mikrokosmos [nos.113, 69, 135, 123, 127, 145, 146], 2 pf, 1940 (B 1947)

115a Suite, 2 pf [free arr. of orch work, Sz 34], 1941 (B 1960)

— Drei Lieder, DD 54, 1898: Im wunderschönen Monat Mai (Heine); Nacht am Rheine (K. Siebel); Die Gletscher leuchten im Mondenlicht

— Liebeslieder, DD 62, 1900 (nos.2, 4 Z 1963): Diese Rose pflück ich hier (Lenau); Du meine Liebe, du mein Herz (Rückert); Ich fühle deinen Odem (Lenau); Wic herrlich leuchtet (Goethe); Herr! der du alles wohl gemacht   7, 34

— Four Songs (L. Pósa), DD 67, 1902 (Bárd 1904); Őszi szellő [Autumn breeze]; Még azt vetik a szememre [They are accusing me]; Nincs olyan bú [There is no greater sorrow]; Ejnye. ejnye! [Alas, alas]; no.1 arr. pf. ?1905–7   8, 9, 34

— Est [Evening] (K. Harsányi), DD 73, ?1903 (Z 1963)

— Four Songs, DD 76, 1903, lost

— Székely Folksong: Piros alma leesett a sárba [The red apple has fallen in the mud], DD C8, 1904 (Magyar Lant 1905)

29 Magyar népdalok [Hungarian folksongs], planned ser.1, inc., 1904–5 (no.1 Z 1963): Lekaszálták már a rétet [They have mowed the pasture already]; [Kiss me for I have to leave]; Fehér László lovat lopott [Fehér László stole a horse]; Az egri ménes mind szürke [The horses of Eger are all grey]; Az egri ménes mind szürke; no.2 arr. pf. ?1905–7

32 A kicsi 'tót'-nak [To the little 'tót' (Béla Oláh-Tóth)] (Hung. children's songs), 1905 (no.3 in J. Demény: Bartók Béla levelek, Budapest, 1948, p.206): Álmos vagyok [I am sleepy]; Ejnye, ejnye, nézz csak ide [Oh, oh, look there]; Puha meleg tolla [The little bird]; Bim bam zúg a harang [Bim bam ring the bells]; Esik eső esdegel [The rain is falling]

33 Magyar népdalok [Hungarian folksongs], 1906 (R 1906), rev. 1938 (Rv 1938, Z 1953): 1 Elindultam szép hazámbul [I left my fair homeland]; 2 Általmennék én a Tiszán ladikon [I would cross the Tisza in a boat]; 3a–b Fehér László lopott [Fehér László stole a horse]; 4a (4 in rev.) A gyulai kert alatt [Behind the garden of Gyula]; 4b (5 in rev.) A kertmegi kert alatt [Behind the garden of Kertmeg]; 5 (not in rev.) Ucca, ucca [The street is on fire]; 6 Ablakomba, ablakomba [From my window shone the moonlight]; 7 Száraz ágtól messze virít [From the withered branch]; 8 Végig mentem a tárkányi [I walked to the end]; 9 Nem messze van ide Kis Margitta [Not far from here is Kis Margitta]; 10 Szánt a babám csireg, csörög [My sweetheart   11, 65

Dille (Budapest, 1970); nos.1, 2, 4, 9, 8 rev. 1928 as Five Hungarian Folksongs (Z 1970)   11

33a Magyar népdalok [Hungarian folksongs], ii, 1906 (nos.4, 6, 7, 8, Z 1963): Tiszán innen, Tiszán túl [On this side of the Tisza, on that side of the Tisza]; Erdők, völgyek, szűk ligetek [Woods, valleys, narrow parks]; Olvad a hó [The snow is melting]; Ha bemegyek a csárdába [Down at the tavern]; Fehér László lovat lopott [Fehér László stole a horse]; Megittam a piros bort [My glass is empty]; Ez a kislány gyöngyöt fűz [This maiden threading]; Sej, mikor engem katonának visznek [The young soldier]; Még azt mondják [And they still say]; Kis kece lányom [My dear daughter]; nos.5, 10 arr. pf, Sz 42/ii/28, 42/i/17

33b Two Hungarian Folksongs, ?1906 (no.1 Z 1963, no.2 in Documenta Bartókiana, iv, 1970): Édesanyám rózsafája [My mother's rose tree]; Túl vagy rózsám, túl vagy a Málnás erdején [My sweetheart, you are beyond the Málnás woods]   12

35b Four Slovakian Folksongs, c1907 (nos.1, 3, 4 Z 1963): V tej bystrickej bráne [Roses in the fields]; Pod lipko nad lipko; Dolu dolinámi [Dirge]; Pritel piák [The message]; no.2 lost

59 Nine Romanian Folksongs, 1915, completed from draft by B. Suchoff: I went off to church one day; Ev'ry lad wants me to perish; Woe is me; See the verdant silken tassel; In the village hall; While I still lived with my mother; You are far away from me; Many thoughts have come into mind; Those who have bad luck

61 Öt dal [5 songs], op.15, 1916 (U 1961, 1966): Tavasz: Az én szerelmem [Spring] (K. Gombossy); Nyár: Szomjasan vágyva [Summer] (Gombossy); A vágyak éjjele [Night of desire] (W. Gleiman); Tél: Színes álomban [Winter] (?Gombossy); Ősz: Itt lent a völgyben [Autumn] (Gombossy)   18, 51

63 Öt dal [5 songs], op.16 (Ady), 1916 (U 1923): Három őszi könnycsepp [Autumn tears]; Az őszi lárma [Autumn echoes]; Az ágyam hívogat [Lost content]; Egyedül a tengerrel [Alone with the sea]; Nem mehetek hozzád [I cannot come to you]   18, 20, 51

63a Slovakian Folksong: Kruti Tono vretena [Tony turns his spindle round], ?1916 (Z 1963)

64 Nyolc magyar népdal [8 Hungarian folksongs], nos.1–5 1907–11, nos.6-8 c1917 (U 1922, B 1955): Fekete főd [Black is the   12, 18

earth]; Istenem, Istenem [My God, my God]; Asszonyok, asszonyok [Wives, let me be one of your company]; Annyi bánat [So much sorrow]; Ha kimegyek [If I climb]; Töltik a nagyerdő útját [They are mending the great forest highway]; Eddig való dolgom [Up to now my work]; Olvad a hó [The snow is melting]

78   Falun (Dedinské scény) [Village scenes] (Slovakian trad.), female v, pf, 1924 (U 1927, B 1954): Ej! hrabajže len [Haymaking]; Letia páivy, letia [At the bride's]; A ty Anča krásna [Wedding]; Beli žemi, beli [Lullaby]; Poza bučky, poza peň [Lads' dance]; nos.3–5 arr. female vv, chamber orch, Sz 79

92   Húsz magyar népdal [20 Hungarian folksongs], 1929 (U 1932) vol.i, Szomorú nóták [Sad songs]; 1 A tömlöcben [In prison]; 2 Régi keserves [Old lament]; 3 Bujdosó ének [The fugitive]; 4 Pásztornóta [Herdsman's song]
vol.ii, Táncdalok [Dancing-songs]; 5 Székely lassú [Slow dance]; 6 Székely friss [Fast dance]; 7 Kanásztánc [Swineherd's dance]; 8 Hatforintos nóta [Six-florin dance]
vol.iii, Vegyes dalok [Diverse songs]; 9 Juhászcsúfoló [The shepherd]; 10 Tréfás nóta [Joking song]; 11 Párositó I [Nuptial serenade]; 12 Párositó II [Humorous song]; 13 Pár-ének [Dialogue song]; 14 Panasz [Complaint]; 15 Bordal [Drinking-song]
vol.iv, Új dalok [New style songs]; 16 (i) Allegro: Hej, édesanyám [Oh, my dear mother]; (ii) Più allegro: Erik a ropogós cseresznye [Ripening cherries]; (iii) Moderato: Már Dobozon [Long ago at Doboz]; (iv) Allegretto: Sárga kukoricaszár [Yellow cornstalk]; (v) Allegro non troppo: Búza, búza [Wheat, wheat]

nos.1, 2, 11, 14, 12 orchd, Sz 101

118 Ukrainian Folksong: A férj keserve [The husband's grief], 1945 (in J. Demeny, ed.: Bartók Béla levelei. Budapest, 1951, p.xiv)
Ukrainian Folksongs, cycle, 1945: Ta ne sa mam [I was not alone]; Ne budu ja vodu piti [I shall not drink the water]; Če my chlopci nekopalci [Not in a ditch, lads]; inc.

EDITIONS, ETC

(concert arrs. for pf)

Italian kbd music, i–xi, arr. c1926–8 (Fischer, 1930): i B. Marcello:

22, 54, 59

24

Sonata, Bb; ii M. Rossi: Toccata no.1, C; iii M. Rossi: Toccata no.2, a; iv M. Rossi: Tre correnti; v–viii A. B. Della Ciaia: Sonata, G; Toccata, Canzone, Primo tempo, Secondo tempo; ix G. Frescobaldi: Toccata, G; x G. Frescobaldi: Fuga, g; xi D. Zipoli: Pastorale, C
J. S. Bach: Sonata VI [bwv530], arr. c1930 (Rv 1930)
Purcell: 2 Preludes, arr. c1930 (Delkas 1947)

(instructive edns. of pf works)

J. S. Bach: Well-tempered Clavier, i–iv (1907–8; vols.i–ii, rev. 2/ c1913–14); 12 Easy Pf Pieces (1916, 2/1924 [13 pieces])
Beethoven: 25 Sonatas (1909–12 [opp.101 and 111 unpubd]); 7 Bagatelles op.33, Variations op.34, 'Eroica' Variations op.35, Polonaise op.79, 11 neue Bagatellen op.119 (1910); Ecossaises (1920)
Chopin: Valses 1–14 (1920)
Couperin: 18 Pieces (1924)
Haydn: 19 Sonatas, nos.1–17 (1911–13), nos.18–19 (1920)
Mozart: 20 Sonatas (1910–12), Fantasy K397/385g (1910)
D. Scarlatti: 10 Sonatas, 1920 (1921, 1926)
Schubert: 2 scherzi (1911)
Schumann: Jugendalbum (1911)
Studies by J. B. Duvernoy, S. Heller, L. Köhler (1917–20)

(critical edns.)

Liszt: Musikalische Werke (Leipzig, 1907–36/R): Hungaria, rev. 1911 (i/5); Ungarischer Marsch, Ungarischer Sturmmarsch, orch, 1916 (i/12); Hungarian Rhapsodies 1–19, pf, 1911–17, unpubd (used by Raabe in ii/12)

(miscellanea)

Cadenza for Beethoven: Pf Conc. no.3, 1st movt, DD A2, c1899–1900
Cadenzas for Mozart: 2 Pf Conc. k365/316a, Sz 121, ?c1940
Composition studies: many exercises (see also DD appx-A), choruses (see DD 61a–b, A4), instrumentations (see DD A1, A10, B1)
Arrs.: Rákóczi March, DD C2, 2 pf, 1896; R. Strauss: Don Quixote, 2 frags., DD C5, vn, 1902: L. van Beethoven: Erlkönig, WoO 131, orchd c1905; D. Zipoli: Suite, 2 pf
Unpubd frags.: juvenilia (see DD appx-B); early sketches (see DD appx-E); piece, vc, str; Rhapsody no.3, vn/vc, pf, 1928; 3 orch frags., 1940s; Str Qt no.7, sketch

ed. B. Suchoff: Rumanian Folk Music (The Hague, 1967–75)

⸻: Turkish Folk Music from Asia Minor (Princeton, 1976); also in L. Vikár, ed.: A. A. Saygun: Béla Bartók's Folk Music Research in Turkey (Budapest, 1976)

WRITINGS

COLLECTIONS

ed. B. Szabolcsi and A. Szőllősy: Bartók Béla válogatott zenei írásai (Budapest, 1948)

ed. D. Carpitella: Béla Bartók: Scritti sulla musica popolare (Turin, 1955)

ed. Z. Vancea: Bartók Béla: Însemnări asupra cîntecului popular [Notes on folksong] (Bucharest, 1956)

ed. E. Hykischová: Béla Bartók: Postřehy a názory [Observations and opinions] (Bratislava, 1965)

ed. D. Dille: Béla Bartók: Ethnomusikologische Schriften (Budapest, 1965–8)

ed. A. Szőllősy: Bartók Béla összegyűjtött írásai [Collected writings], i (Budapest, 1966)

ed. B. Suchoff: Béla Bartók: Rumanian Folk Music (The Hague, 1967–75) [see also 'Books']

ed. B. Szabolcsi: Béla Bartók: Musiksprachen, Aufsätze und Vorträge (Leipzig, 1972)

ed. B. Suchoff: Béla Bartók Essays (London, 1976) [review by L. Somfai in MT, cxviii (1977), 395]

⸻: Yugoslav Folk Music (Albany, 1978)

BOOKS

Cântece populare românești din comitatul Bihor (Ungaria)/Chansons populaires roumaines du département Bihar (Hongrie) (Bucharest, 1913); repr. in Dille (1965–8)

Volksmusik der Rumänen von Maramureş (Munich, 1923); repr. in Dille (1965–8); Eng. trans. in Suchoff (1967–75), vol.v

A magyar népdal [Hungarian folksong] (Budapest, 1924; Ger. trans., 1925; Eng. trans., 1931); repr. in Hung. and Ger. in Dille (1965–8); repr. in Eng. in The Hungarian Folk Song, ed. B. Suchoff (Albany, 1981)

Melodien der rumänischen Colinde (Weihnachtslieder) (Vienna, 1935); repr. with unpubd pt.2 in Dille (1965–8); Eng. trans. in Suchoff (1967–75), vol.iv

with A. B. Lord: Serbo-Croatian Folk Songs (New York, 1951)

ed. A. B. Lord: Serbocroatian Heroic Songs, i (Cambridge, Mass., and Belgrade, 1954); repr. in Suchoff (1978), vol.i

Slovenské l'udove piesne/Slowakische Volkslieder, i–ii (Bratislava, 1959–70) [vol.iii unpubd]

ARTICLES

* – repr. in A. Szőllősy, ed.: Bartók Béla összegyűjtött írásai, i (Budapest, 1966)

E – repr./no. in Béla Bartók Essays, ed. B. Suchoff (London, 1976) [ed. E – version or compilation]

(on himself and his music)

'Kossuth': szinfóniai költemény: irta Bartók Béla', Zeneközlöny, ii (1904), 82*; Eng. trans, Hallé Concert Society programme (18 Feb 1904), 506 [E 48]

'Szvit nagy zenekarra (op.3)', Zeneközlöny, vii (1908)

'II. suite', Zeneközlöny, viii (1909)

'Rhapsodie für Klavier und Orchester (op.1) von Béla Bartók', Die Musik, ix (1909–10), 226* [E 49]

'Bartók Béla', Budapesti újságírók egyesülete almanachja (1911), 292*

'"A fából faragott királyfi"': a M. Kir. Operaház bemutatójához, ii: a zeneszerző a darabjáról', Magyar színpad, xx/105 (1917), 2* [on the première of The Wooden Prince] [E 50]

'"A Kékszakállú herceg vára": az Operaház újdonsága, i: szerzők a darabjukról', Magyar színpad, xxi/143 (1918), 1* [on Bluebeard's Castle] [E 51]

'Béla Bartók (Selbstbiographie)', Musikpädagogische Zeitschrift, viii/11–12 (1918), 97*; rev. in Rheinische Musik und Theaterzeitung, xx/5–6 (1919), 1; Musikblätter des Anbruch, iii (1921), 87[E 52]; Magyar írás, i/2 (1921), 33; Az est hármaskönyve (Budapest, 1923), 77; Sovremennaya muzïka, ii/7 (1925), 1; Színházi élet, xvii/51 (1927), 49; collected in Documenta Bartókiana, ii (1965), 109

Suppl. to programme for Liverpool pf recital (30 March 1922)

Preface to score of Fourth Quartet (Vienna, 1930)* [E 53]

Suppl. to programme for first concert of New Hungarian Musical Society (20 Jan 1932)* [on the 44 Duos]

Analysis of Fifth Quartet, ed. L. Somfai, Muzsika, xiv/12 (1971), 26 [E 54]

Preface to score of Music for Strings, Percussion and Celesta (Vienna, 1937)* [E 55]

25, 30   17   17   22   22, 24   22

'Béla Bartók über sein neuestes Werk', *National Zeitung* (Basle, 13 Jan 1938)* [on Sonata for Two Pianos and Percussion] [E 56]

Suppl. to programme for *III. Internationalen zeitgenössisches Musikfest: Baden-Baden 1938*, 23 [on 5 Hungarian Folksongs Sz 101]

'Béla Bartók à l'Orchestre de la Suisse romande: analyse du "Deuxième concerto" pour piano et orchestre de Béla Bartók par son auteur', *La radio* (Lausanne, 17 Feb 1939), 280, 282* [ed. E 57] [see Somfai, *New Hungarian Quarterly* (1981)]

1940 lecture, 'Contemporary Music in Piano Teaching' [on *For Children* and *Mikrokosmos*] [E 59]

1941 Columbia lecture, 'The Relation between Contemporary Hungarian Art Music and Folk Music' [E 45]

1943 Harvard lectures [E 46] [on 'new' Hungarian music and Bartók's musical language; see also Somfai, *MT*, cxviii (1977), 395]

'Ask the Composer' [recorded interview, 2 July 1944, on pf pieces], ed. G. Kroó, *SM*, xi (1969), 253; in Bartók Record Archives (1981), 10/4

Note on Concerto for Orchestra, *Concert Programme of the Boston Symphony Orchestra* (1944–5), no.8, pp.442, 444* [E 60]

'I salute the valiant Belgian people', *Belgium*, v (1945), 563 [E 62]

'Béla Bartók', facs. in *10 Easy Pieces*, pf (London, c1950) [see also T. Tallián: 'Bartók marginália', *Zenetudományi dolgozatok 1979* (Budapest, 1979), 36; It. version, *Musica-realtà*, ii/5 (1981), 173]

(others)

'Strauss: Sinfonia domestica (op.53)', *Zeneközlöny*, iii (1905), 137* [E 63]

'Székely balladák', *Ethnographia*, xix (1908), 43, 105*

'Dunántúli balladák' [Transdanubian ballads], *Ethnographia*, xx (1909), 301*

'Strauss: Elektra', *A zene*, ii/4 (1910), 57* [E 64]

'A hangszeres zene folklorêja Magyarországon' [Instrumental musical folklore in Hungary], *Zeneközlöny*, ix (1911), 141, 207, 309; x (1912), 601* [ed. E 33]

'A magyar zenéről', *Aurora*, i (1911), 126*; Ger. trans., *Der Merker*, vii (1916), 757 [E 38]

'A magyar nép hangszerei' [Hungarian folk instruments], *Ethnographia*, xxii (1911), 305; xxiii (1912), 110 [ed. E 33]

'Delius-bemutató Bécsben' [A Delius première in Vienna], *Zeneközlöny*, ix (1911), 340* [E 66]

'Liszt zenéje és a mai közönség' [Liszt's music and today's public], *Népművelés*, vi (1911), 359* [E 67]

'A clavecinre irt művek előadása' [The performance of works written for the harpsichord], *Zeneközlöny*, x (1912), 226* [E 34]

'Az összehasonlító zenefolklore' [Comparative musical folklore], *Ujélet népművelés*, i (1912), 109* [E 22]

'A hunyadi román nép zenedialektusa' [The folk music dialect of the Hunedoara Romanians], *Ethnographia*, xxv (1914), 108*; Ger. trans., *ZMw*, ii (1919–20), 352 [E 14]

'Observări despre muzica poporală românească', *Convorbiri literare*, xlviii (1914), 703* [E 26]

'A Biskra-vidéki arabok népzenéje' [The folk music of the Arabs of the Biskra region], *Szimfónia*, i (1917), 308*; Ger. trans., *ZMw*, ii (1919–20), 489–522

'Primitiv népi hangszerek Magyarországon' [Primitive folk instruments in Hungary], *Zenei szemle*, i (1917), 273, 311* [ed. E 33]

'A magyar népzenéről' [Hungarian folk music], draft, before 1918, ed. T. Tallián: 'Bartók marginália', *Zenetudományi dolgozatok 1979* (Budapest, 1979), 36

'Die Melodien der madjarischen Soldatenlieder', *K. u. K. Kriegsministerium Musikhistorische Zentrale: Historisches Konzert am 12. Jänner 1918* (Vienna, 1918), 36* [E 10]

'A népzene (paraszt-zene) fejlődési fokai' [Stages in the development of folk music (peasant music)] before 1920, *Documenta Bartókiana*, iv (1970), 88

'Musikfolklore', *Musikblätter des Anbruch*, i (1919), 102* [E 23]

'Das Problem der neuen Musik', *Melos*, i (1920), 107* [E 68] [other Ger. and Hung. versions in *Documenta Bartókiana*, v (1977), 23]

'Hungary in the Throes of Reaction', *Musical Courier*, no.80 (20 April 1920), 42 [E 69]; Ger. draft, *Documenta Bartókiana*, v (1977), 33

'The Peasant Music of Hungary', 1920, *Musical Courier*, no.103/11 (12 Sept 1931), 6; Ger. draft in *Documenta Bartókiana*, iv (1970), 101

'Slovak Peasant Music', 1920, *Musical Courier*, no.103/13 (26 Sept 1931), 6 [E 19]

[About Romanian peasant music], Ger. draft, c1920, *Documenta Bartókiana*, iv (1970), 107

'Bartók válasza Hubay Jenőnek' [Reply to Hubay], *Szózat*, ii/125 (1920), 2* [E 27]

'Hungarian Peasant Music', c1920, *MQ*, xix (1933), 267* [E 13]

'Kodály's new Trio a Sensation Abroad', *Musical Courier*, no.81 (19 Aug 1920), 5 [E 70]

'Die Volksmusik der Völker Ungarns', Ger. draft, c1920–21, *Documenta Bartókiana*, iv (1970), 112

'Der Einfluss der Volksmusik auf die heutige Kunstmusik', *Melos*, i (1920), 384* [E 40]

'Arnold Schönbergs Musik in Ungarn', *Musikblätter des Anbruch*, ii (1920), 647* [E 71]

'To Celebrate the Birth of the Great Bonn Composer', *Musical Courier*, no.81 (23 Dec 1920), 7; in *Documenta Bartókiana*, v (1977), 54

'Kodály Zoltán', *Nyugat*, xiv (Budapest, 1921), 235* [E 72]

'Schönberg and Stravinsky enter "Christian–National" Budapest without Bloodshed', *Musical Courier*, no.82 (21 Feb 1921), 7; repr. in *Documenta Bartókiana*, v (1977), 65

'New Kodály Work raises Storm of Critical Protest', *Musical Courier*, no.82 (31 March 1921), 6; repr. in *Documenta Bartókiana*, v (1977), 73

'Lettera da Budapest I', *Il pianoforte*, ii/5 (1921), 153* [E 73]

'Budapest sorely misses Dohnányi', *Musical Courier*, no.82 (26 May 1921), 47; repr. in *Documenta Bartókiana*, v (1977), 85

'The Relation of Folk-song to the Development of the Art Music of our Time', *The Sackbut*, ii (1921), 5* [E 41]

'Aki nem tud arabusul' [He who knows no Arabic], *Szózat*, iii/32 (1921), 4* [E 28]

'Budapest welcomes Dohnányi's Return', *Musical Courier*, no.83 (14 July 1921), 37; repr. in *Documenta Bartókiana*, v (1977), 108

'Lettera da Budapest II', *Il pianoforte*, ii/9 (1921), 277* [E 75]

'Della musica moderna in Ungheria', *Il pianoforte*, ii/7 (1921), 193* [E 74]; in Eng. as 'The Development of Art Music in Hungary', *The Chesterian*, no.20 (1922), 101; Hung. and Ger. drafts, *Documenta Bartókiana*, v (1977), 114

'Two Unpublished Liszt Letters to Mosonyi', *ReM*, ii/1 (1921), 8* [E 11]

'La musique populaire hongroise', *ReM* etc., *A Dictionary of Modern Music and Musicians* (London, 1924/R1971)

draft, ed. L. Somfai, *Magyar zene*, xvi (1975), 115

'Zenefolklore-kutatások Magyarországon' [Research on Hungarian musical folklore], *Zenei szemle*, xiii (1929), 13*; rev. Fr. trans. in *Ier Congrès international des arts populaires: Prague 1928*, ii, 127 [E 24]

'Magyar népi hangszerek', 'Román népzene', 'Szlovák népzene' [Hungarian folk instruments; Romanian folk music; Slovak folk music], *Zenei lexikon* (Budapest, 1930–31)*[ed. E 33, E 15, 18]

'Cigányzene? Magyar zene? [Magyar népdalok a német zenemüpiacon) [Gypsy music? Magyar music? (On the edition of Hungarian folksongs)], *Ethnographia*, xiii (1931), 62*; Ger. trans., *Ungarische Jb*, xi (1931), 191 [E 29]

Möller, Heinrich: über die Herausgabe ungarischer Volkslieder', *ZMw*, xiii (1930–31), 580* [book review]

'Nochmals: über die Herausgabe ungarischer Volkslieder', *ZMw*, xiv (1931–2), 179*

'Gegenantwort an Heinrich Möller', *Ungarische Jb*, xii (1932), 130*

'Mi a népzene?', 'A parasztzene hatása az újabb müzenére', 'A népzene jelentőségéről' [What is folk music? The influence of peasant music on contemporary composition; On the significance of folk music], *Uj idők*, xxxvii (1931), 626, 718, 818* [E 2, 43, 44]; Ger. trans., abridged, *Mitteilungen der Österreichischen Musiklehrerschaft* (1932), no.2, p.6; no.3, p.5

'Neue Ergebnisse der Volksliedforschung in Ungarn', *Anbruch: Monatsschrift für Moderne Musik*, xiv (1932), 37*

'Proposition de M. Bela Bartok concernant les éditions de textes authentiques (Urtextausgaben) des œuvres musicales', League of Nations, Commission de coopération intellectuelle, Souscommission arts et lettres 1931–1938 [minutes]* [E 82]

'Ungarische Volksmusik', *Schweizerische Sängerzeitung*, xxxiii (1933), 13, 21, 31* [E 12]

[Proposals on Arab folk music], orig. Fr., *Documenta Bartókiana*, iv (1970), 117

'Zum Kongress für arabische Musik – Kairo 1932', *Zeitschrift für vergleichende Musikwissenschaft*, i/2 (1933), 46* [E 8]

'Rumänische Volksmusik', *Schweizerische Sängerzeitung*, xxxiii (1933), 141, 148, 168* [E 16]

'Staat und Kunst', Ger. draft, c1934, ed. T. Tallián, Arion 13 (Budapest, 1982), 104

'Béla Bartók replies to Percy Grainger', Music News (19 Jan 1934), 9* [E 30]

Népzenénk és a szomszéd népek népzenéje [The folk music of the Magyars and neighbouring peoples] (Budapest, 1934]*; Ger. trans., Ungarische Jb, xv (1935), 194–258; Fr. trans., Archivum europae centro-orientalis, ii (1936), 197–232 and i–xxxii

'Miért gyűjtsünk népzenét?' [Why do we collect folk music?], Zeneművészeti főiskola évkönyve (Budapest, 1935), 3*

'Magyar népzene', 'Román népzene', 'Szlovák népzene' [Hungarian, Romanian, Slovak folk music], Révai nagy lexikona, xxi (Budapest, 1935), 571, 725, 776*

'Nachwort zu dem "Volksmusik der Magyaren und der benachbarten Völker" – Antwort auf einen rumänischen Angriff', Ungarische Jb, xvi (1936), 233; Hung. orig., Szép szó, ii (1937), 263 [E 32]

'Liszt Ferenc', Nyugat, xxix (1936), 171*; Fr. trans., ReM (1936), no.167, p.1 [E 83]

'Musique et chanson populaires', AcM, viii (1936), 97*; Hung. orig., Szép szó, i (1936), 274

Miért és hogyan gyűjtsünk népzenét? A zenei folklore törvénykönyve [Why and how do we collect folk music?] (Budapest, 1936]* [E 3]

'A gépzene' [Mechanical music], Szép szó, ii/11 (1937), 1* [E 37]

'Vorschläge für die Einrichtung eines Volksmusik-Archivs', Documenta Bartókiana, iv (1970), 124

'Zenére való nevelés' [On music education], ed. L. Somfai, Muzsika, xiv/4 (1971), 1 [E 84]

'Népdalgyüjtés Törökországban' [Collecting folksongs in Anatolia], Nyugat, i (1937), 173*; Eng. trans., Hungarian Quarterly, iii (1937), 337 [E 20]

'Népdalkutatás és nacionalizmus' [Folksong research and nationalism], Tükör, v (1937), 166*; Fr. and Ger. trans., Revue internationale de musique, i (1938), 608

'Erklärung', 28 March 1938, Documenta Bartókiana, iv (1970), 148

'Az úgynevezett bolgár ritmus' [The so-called Bulgarian rhythm], Énekszó, v (1937–8), 537*

'Opinion de M. Béla Bartók (Varsovie) [!]', Revue internationale de musique, i (1938), 452* [E 85]

[Ravel], ReM, xix (1938), no.187, p.436* [E 86]

[Some Problems of Folk Music Research in East Europe], Harvard lecture, 23 April 1940, pre 1941 [E 25] (orig. in Ger.)

'Race Purity in Music', MM, xix/3–4 (1941–2), 153*

[On American and British folk music material], 1 June 1942 [E 7]

'Parry Collection of Yugoslav Folk Music', New York Times (28 June 1942)*

'Diversity of Material Yielded up in Profusion in European Melting-pot' [MS title: 'Folk Song Research in Eastern Europe'], Musical America, lxiii/1 (10 Jan 1943), 27* [E 6]

'Szabolcsi, Bence', Universal Jewish Encyclopedia, x (New York, 1943), 138* [E 87]

'To Sir Henry Wood', Homage to Sir Henry Wood (London, 1943), 36 [E 88]

'Hungarian Music', American Hungarian Observer (New York, June 1944), 3* [E 47]

'Some Linguistic Observations', Tempo, no.14 (1946), 5* [E 89]

FOLKSONG EDITIONS AND TRANSCRIPTIONS

with Z. Kodály: Erdélyi magyarság: népdalok [Transylvanian Hungarian folksongs], Hung., Eng., Fr. edns. (Budapest, 1923)

'Musique paysanne serbe et bulgare du Banat', Budapest 1935, Documenta Bartókiana, iv (1970), 221

with Z. Kodály: Magyar népzenei gramofonfelvételek: 1. sorozat [Hungarian folk music gramophone recordings: 1st series] (Budapest, 1937)

with J. Deutsch and S. Veress: A Magyar Rádió és a Néprajzi Múzeum gyüjteménye [Gramophone record collection of the Hungarian Radio and of the Ethnographic Museum] (Budapest, 1937–9); ed. L. Somfai in Hungarian Folk Music: Gramophone Records with Bartók's Transcriptions (Budapest, 1981)

[50 csángómagyar népdal] [50 csángó-Hungarian folksongs] in P. P. Domokos: Mert akkor az idő napkeletre fordul (Cluj, 1940)

with Z. Kodály: Corpus musicae popularis hungaricae (Budapest, 1951–)

Magyar népdalok: egyetemes gyüjtemény [Hungarian folksongs: complete edition] (Budapest, in preparation)

RECORDINGS

Centenary Edition of Bartók's Records (Complete), i: Bartók at the Piano 1920–1945; ii: Bartók Record Archives: Bartók plays and talks 1912–1944, ed. L. Somfai, Z. Kocsis, J. Sebestyén (Budapest, 1981)

BIBLIOGRAPHY

DOCUMENTS AND CATALOGUES

J. Demény, ed.: *Bartók Béla levelei* [Letters] (Budapest, 1948–71, enlarged 2/1976; Ger. trans., 1960, enlarged 2/1973; Eng. trans., 1971)

——: 'Bartók Béla tanuló évei és romantikus korszaka' [Bartók's years of study and his romantic period], *Zenetudományi tanulmányok*, ii (1954) [documents of 1899–1903]

——: 'Bartók Béla művészi kibontakozásának évei: találkozás a népzenével' [Bartók's years of artistic development: contact with folk music], *Zenetudományi tanulmányok*, iii (1955) [documents of 1906–14]

B. Szabolcsi, ed.: *Bartók: sa vie et son oeuvre* (Budapest, 1956, rev. 2/1968; Ger. trans., 1957, enlarged 2/1972) [incl. catalogue of works and writings compiled by Szőllősy]

J. Ujfalussy, ed.: *Bartók breviárium* (Budapest, 1958, rev. 2/1974, 3/1980; Swed. edn., 1981) [letters, essays and documents]

W. Reich, ed.: *Béla Bartók: eigene Schriften und Erinnerungen der Freunde* (Basle, 1958)

J. Demény, ed.: 'Bartók Béla megjelenése az európai zeneéletben' [Bartók's appearance in European musical life], *Zenetudományi tanulmányok*, vii (1959) [documents of 1914–26]

——: 'Bartók Béla pályája delelőjén' [The zenith of Bartók's career], *Zenetudományi tanulmányok*, x (1962) [documents of 1926–40]

V. Bator: *The Béla Bartók Archives: History and Catalogue* (New York, 1963)

F. Bónis: *Béla Bartóks Leben in Bildern* (Budapest, 1964, enlarged 2/1972 in Hung. and Ger.; Eng. trans., 1972)

D. Dille, ed.: *Documenta Bartókiana*, i–iv (Budapest, 1964–70)

E. Helm: *Béla Bartók in Selbstzeugnissen und Bilddokumenten* (Hamburg, 1965)

ed. J. Vinton: 'Bartók on his own Music', *JAMS*, xix (1966), 232

V. Čižik, ed.: *Bartóks Briefe in die Slowakei* (Bratislava, 1971)

D. Dille: *Thematisches Verzeichnis der Jugendwerke Béla Bartóks 1890–1904* (Budapest, 1974 [review by T. Tallián in *SM*, xvii (1975), 427])

F. László, ed.: *Béla Bartók scrisori I–II* [Letters] (Bucharest, 1976)

L. Somfai, ed.: *Documenta Bartókiana*, v– (Budapest, 1977–)

V. Lampert: *Bartók népdalfeldolgozásainak forrásjegyzéke* [Thematic catalogue of the sources of Bartók's folksong arrangements] (Budapest, 1980), Ger. trans., *Documenta Bartókiana*, vi (1982)

Ph. A. Autexier, ed.: *Béla Bartók: musique de la vie* (Paris, 1981)

B. Bartók jr: *Apám életének krónikája* [Chronicle of my father's life] (Budapest, 1981]

———: *Bartók Béla családi levelei* [Bartók's family letters] (Budapest, 1981)

———: *Bartók Béla műhelyében* [In Bartók's workshop] (Budapest, 1982)

MONOGRAPHS AND COLLECTIONS OF ESSAYS

A. Molnár: *Bartók Két elégiájának elemzése* [Analysis of Bartók's Two Elegies] (Budapest, 1921)

E. von der Null: *Béla Bartók: ein Beitrag zur Morphologie der neuen Musik* (Halle, 1930)

E. Haraszti: *Béla Bartók: his Life and Works* (Paris, 1938)

D. Dille: *Béla Bartók* (Antwerp, 1939)

M. Seiber: *The String Quartets of Béla Bartók* (London, 1945; Ger. trans., 1945)

S. Moreux: *Béla Bartók: sa vie, ses oeuvres, son langage* (Paris, 1949; Ger. trans., 1950; Eng. trans., 1953)

Tempo, nos.13–14 (1949–50) [special issue]

H. and J. Geraedts: *Béla Bartók* (Harlem, 1951, 2/1961)

H. U. Engelmann: *Béla Bartóks Mikrokosmos: Versuch einer Typologie 'Neuer Musik'* (Würzburg, 1953)

H. Stevens: *The Life and Music of Béla Bartók* (New York, 1953, rev. 2/1964)

Musik der Zeit (1953), no.3 [special issue]

J. Uhde: *Bartóks Mikrokosmos: Spielanweisungen und Erläuterungen* (Regensburg, 1954)

Musik der Zeit (1954), no.9 [special issue]

E. Lendvai: *Bartók stílusa* (Budapest, 1955) [analysis of Sonata for Two Pianos and Percussion and Music for Strings, Percussion and Celesta]

ReM (1955), no.224 [special issue]

Z. Kodály and others, eds.: *Studia memoriae Belae Bartók sacra* (Budapest, 1956, 3/1959)

J. Szegő: *Bartók Béla, a népdalkutató* [Bartók the folklorist] (Bucharest, 1956)

R. Traimer: *Béla Bartóks Kompositionstechnik dargestellt an seinen sechs Streichquartetten* (Regensburg, 1956)

B. Suchoff: *Guide to Bartók's Mikrokosmos* (London, 1957, rev. 2/1971)

A. Fassett: *Béla Bartók's American Years: the Naked Face of Genius* (Boston, Mass., 1958)

L. Lesznai: *Béla Bartók: sein Leben, seine Werke* (Leipzig, 1961; Eng. trans., rev., 1973)

B. Szabolcsi: *Béla Bartók: Leben und Werk* (Leipzig, 1961, enlarged 2/1968)

Bibliography

Liszt–Bartók: 2nd International Conference: Budapest 1961 [*SM*, v (1963)]

F. Fricsay: *Über Mozart und Bartók* (Copenhagen, 1962)

G. Kroó: *Bartók színpadi müvei* [Bartók's stage works] (Budapest, 1962)

Zenetudományi tanulmányok, x (1962) [special issue]

P. Citron: *Bartók* (Paris, 1963)

Z. Pálová-Vrbová: *Béla Bartók: život a dílo* [Bartók: life and works] (Prague, 1963)

E. Lendvai: *Bartók dramaturgiája* (Budapest, 1964)

W. Rudziński: *Warsztat kompozytorski Béli Bartóka* [Bartók's compositional workshop] (Kraków, 1964)

P. Meyer: *Béla Bartóks 'Ady-Lieder' op.16* (diss., U. of Zurich, 1965)

J. Ujfalussy: *Bartók Béla* (Budapest, 1965, rev. 2/1970, 3/1976; Eng. trans., 1971; Russ. trans., 1971; Ger. trans., 1973)

J. W. Downey: *La musique populaire dans l'oeuvre de Béla Bartók* (diss., U. of Paris, 1966)

J. Kárpáti: *Bartók vonósnégyesei* [Bartók's string quartets] (Budapest, 1967, rev., enlarged 2/1976 as *Bartók kamarazenéje* [Bartók's chamber music]; Eng. trans., 1975)

J. Demény: *Bartók Béla a zongoraművész* [Bartók the pianist] (Budapest, 1968, 2/1973)

I. Martïnov: *Bela Bartok* (Moscow, 1968)

E. Kapst: *Die 'polymodale Chromatik' Béla Bartóks: ein Beitrag zur stilkritischen Analyse* (diss., U. of Leipzig, 1969); abridged in *BMw*, xxi (1970), 1

I. Nest'yev: *Bela Bartok 1881–1945: zhizn' i tvorchestvo* [Bartók: life and works] (Moscow, 1969)

A. Benkő: *Bartók Béla romániai hangversenyei* [Bartók's concerts in Romania] (Bucharest, 1970)

P. Petersen: *Die Tonalität im Instrumentalschaffen von Béla Bartók* (Hamburg, 1971)

T. A. Zieliński: *Bartók* (Kraków, 1969; Ger. trans., 1973)

A. Benkő: *Bartók Béla romániai hangversenyei 1922–1936* [Bartók's concerts in Romania 1922–36] (Bucharest, 1970)

G. Weiss: *Die frühe Schaffensentwicklung Béla Bartóks im Lichte westlicher und östlicher Traditionen* (diss., U. of Erlangen-Nuremberg, 1970)

T. Hundt: *Bartók's Satztechnik in den Klavierwerken* (Regensburg, 1971)

G. Kroó: *Bartók kalauz* [A guide to Bartók] (Budapest, 1971; Eng. trans., 1974; Ger. trans., 1974)

E. Lendvai: *Bartók költői világa* [Bartók's poetic world] (Budapest, 1971)

——: *Béla Bartók: an Analysis of his Music* (London, 1971)

International Musicological Conference in Commemoration of Béla Bartók: Budapest 1971

O. Nordwall: *Béla Bartók: Traditionalist-modernist* (Stockholm, 1972)

F. Bónis, ed.: *Magyar zenetörténeti tanulmányok*, iii: *Mosonyi Mihály és Bartók Béla emlékére* [Hungarian studies in musicology, iii: in memory of Mosonyi and Bartók] (Budapest, 1973)

W. Fuchss: *Béla Bartók und die Schweiz* (Berne, 1973)

F. László, ed.: *Bartók-dolgozatok* [Bartók studies] (Bucharest, 1974)

J. McCabe: *Bartók Orchestral Music* (London, 1974)

H. Fladt: *Zur Problematik traditioneller Formtypen dargestellt an Sonatensätzen in den Streichquartetten Béla Bartóks* (Munich, 1974)

E. Lendvai: *Bartók és Kodály harmóniavilága* [The system of harmony of Bartók and Kodály] (Budapest, 1975)

E. Antokoletz: *Principles of Pitch Organization in Bartók's Fourth String Quartet* (diss., City U. of New York, 1975)

V. Lampert: *Bartók Béla: a múlt magyar tudósai* [Bartók as a Hungarian scholar] (Budapest, 1976)

Y. Lenoir: *Vie et oeuvre de Béla Bartók aux États-Unis d'Amérique 1940–1945* (diss., U. of Louvain, 1976)

F. Michael: *Béla Bartóks Variationstechnik dargestellt im Rahmen einer Analyse seines 2. Violinkonzert* (Regensburg, 1976)

A. Szentkirályi: *Bartók's Second Sonata for Violin and Piano* (diss., Princeton U., 1976)

T. Crow, ed.: *Bartók Studies* (Detroit, 1976)

L. Vikár, ed.: *A. A. Saygun: Béla Bartók's Folk Music Research in Turkey* (Budapest, 1976)

D. Dille: *Généalogie sommaire de la famille Bartók* (Antwerp, 1977)

I. Oramo: *Modaalinen symmetria: tutkimus Bartókin kromatiikasta* (Helsinki, 1977)

D. Dille: *Het werk van Béla Bartók* (Antwerp, 1979)

E. Lendvai: *Bartók and Kodály* (Budapest, 1979)

J. Gergely: *Bela Bartok: compositeur hongrois, ReM* (1980), nos.328–35

F. László: *Bartók Béla: tanulmányok és tanúságok* [Bartók: studies and conclusions] (Bucharest, 1980)

T. Tallián: *Bartók Béla* (Budapest, 1981)

Y. Queffélec: *Béla Bartók* (Paris, 1981)

F. Bónis, ed.: *Így láttuk Bartókot: harminchat emlékezés* [As we saw Bartók: 36 recollections] (Budapest, 1981)

L. Somfai: *Tizennyolc Bartók-tanulmány* [18 Bartók studies] (Budapest, 1981)

F. László, ed.: *Bartók-dolgozatok 1981* [Bartók studies 1981] (Bucharest, 1982)

Bibliography

L. Somfai, ed.: report of the International Bartók Symposium Budapest 1981, *SM*, xxiv/3–4 (1982)

T. Tallián, ed.: 'Bartók and Words', *Arion 13*, ed. G. Somlyó (Budapest, 1982), 67–181

E. Antokoletz: *The Works of Béla Bartók: a Study of Tonality and Progression in Twentieth-Century Music* (Berkeley, 1983)

T. Tallián: *Cantata profana – az átmenet mitosza* [– The myth of metamorphosis] (Budapest, 1983)

OTHER LITERATURE

E. Wellesz: 'Ungarische Musik, i: Béla Bartók', *Musikblätter des Anbruch*, ii (1920), 225

Z. Kodály: 'Bela Bartok', *ReM*, ii/5 (1921), 205

M. D. Calvocoressi: 'Béla Bartók: an Introduction', *MMR*, lii (1922), 54

H. Leichtentritt: 'On the Art of Béla Bartók, *MM*, vi/3 (1928–9), 3

S. Jemnitz: 'Béla Bartók, v: Streichquartett', *IMSCR*, iii *Barcelona 1936*, 127

H. Pleasants and T. Serly: 'Bartók's Historic Contribution', *MM*, xvii (1940), 131

G. Abraham: 'The Bartók of the Quartets', *ML*, xxvi (1945), 185

P. Sacher: 'Béla Bartók zum Gedächtnis', *Mitteilungen des Baseler Kammerorchesters* (1945), no.12; repr. in *Musik der Zeit* (1953), no.3, p.66

R. Leibowitz: 'Béla Bartók ou la possibilité de compromis dans la musique contemporaine', *Les temps modernes* (Paris, 1947), 705

E. Lendvai: 'Bartók: Az éjszaka zenéje' [Bartók: The Night Music], *Zenei szemle*, i (1947), 216

J. Szigeti: *With Strings Attached* (New York, 1947, rev. 2/1967)

C. Brăiloiu: 'Béla Bartók folkloriste', *SMz*, lxxxviii (1948), 92

M. Babbitt: 'The String Quartets of Bartók', *MQ*, xxxv (1949), 377

Z. Kodály: 'A folklorista Bartók', *Új zenei szemle*, i (1950), 33; Ger. trans., *Musik der Zeit*, (1954), no.9, p.33

C. Mason: 'Béla Bartók and Folksong', *MR*, xi (1950), 292

B. Szabolcsi: 'Bartók és a népzene' [Bartók and folk music], *Új zenei szemle*, i (1950), 39; Fr. trans. in Szabolcsi (1956), 75

L. Hernádi: 'Bartók Béla, a zongoraművész, a pedagógus, az ember' [Bartók the pianist, the teacher, the man], *Új zenei szemle*, iv (1953), 1; Fr. trans., *ReM* (1955), no.224, p.77

B. Bartók jr: 'Apámról' [About my father], *Zenetudományi tanulmányok*, iii (1955), 281

Z. Kodály: 'Szentirmaytól Bartókig' [From Szentirmay to Bartók], *Új zenei szemle*, vi (1955), 6

B. Szabolcsi: 'Bartók Béla élete' [Bartók's life], *Csillag*, ix (1955), 1855; Fr. trans. in Szabolcsi (1956), 9–45

———: 'A csodálatos mandarin', *Zenetudományi tanulmányok*, iii (1955), 519; Fr. trans., *SM*, i (1961), 341

C. Mason: 'An Essay in Analysis: Tonality, Symmetry and Latent Serialism in Bartók's Fourth Quartet', *MR*, xviii (1957), 189

B. Szabolcsi: 'Bartók és a világirodalom' [Bartók and world literature], *Nagyvilág*, iv (1959), 265; Fr. trans., *Europe*, xli (1963), 221

A. Forte: 'Bartók's Serial Composition', *MQ*, xlvi (1960), 233

G. Kroó: 'Duke Bluebeard's Castle', *SM*, i (1961), 251–340

E. Lendvai: 'Der wunderbare Mandarin', *SM*, i (1961), 363–432

J. Ujfalussy: 'A híd-szerkezetek néhány tartalmi kérdése Bartók müvészetében' [Some inherent questions of arch symmetry in Bartók's works], *Zenetudományi tanulmányok*, x (1962), 15; Ger. trans., *SM*, v (1963), 541

F. Bónis: 'Quotations in Bartók's Music', *SM*, v (1963), 355

J. Kárpáti: 'Béla Bartók and the East', *SM*, vi (1964), 179

I. Waldbauer: 'Bartók's First Piano Concerto: a Publication History', *MQ*, li (1965), 336

J. Chailley: 'Essai d'analyse du Mandarin merveilleux', *SM*, viii (1966), 11

J. Vinton: 'Toward a Chronology of the Mikrokosmos', *SM*, viii (1966), 41

G. French: 'Continuity and Discontinuity in Bartók's "Concerto for Orchestra" ', *MR*, xxviii (1967), 122

A. Mihály: 'Metrika Bartók IV. vonósnégyesének II. tételében' [Metrical concept of the 2nd movement of Bartók's 4th Quartet], *Muzsika*, x (1967), no.10, p.18; no.11, p.34; no.12, p.35

W. Pütz: *Studien zum Streichquartettschaffen bei Hindemith, Bartók, Schönberg und Webern* (Regensburg, 1968)

B. Suchoff: 'Structure and Concept in Bartók's Sixth Quartet', *Tempo*, no.83 (1967–8), 2

F. Bónis, ed.: *A. Tóth: Válogatott zenekritikái 1934–1939* [Selected criticism 1934–9] (Budapest, 1968)

I. Szelényi: *A népdalharmonizálás alapelvei Bartók 'Gyermekeknek' címü müve alapján* [Theory of folksong harmonization, based on Bartók's *For Children*] (Budapest, 1968)

L. Bárdos: 'Népi ritmusok Bartók müzenéjében' [Folksong rhythm in Bartók's compositions], *Harminc irás* (Budapest, 1969), 9

J. Kárpáti: 'Les gammes populaires et le système chromatique dans l'oeuvre de Béla Bartók', *SM*, xi (1969), 227

G. Kroó: 'Bartók Béla megvalósulatlan kompozíciós terveiről' [Unrealized plans and ideas for projects by Bartók], *Magyar zene*, x (1969), 251; Eng. trans., *SM*, xii (1970), 11

Bibliography

——'Adatok "A kékszakállú herceg vára" keletkezéstörténetéhez' [Some data on the genesis of *Duke Bluebeard's Castle*], *Magyar zenetörténeti tanulmányok* (1969), 333

L. Somfai: ' "Per finire": some Aspects of the Finale in Bartók's Cyclic Form', *SM*, xi (1969), 391

E. Kapst: 'Stilkriterien der polymodal-kromatischen Gestaltungsweise im Werk Béla Bartóks', *BMw*, xii (1970), 1

K. Stockhausen: 'Bartók's Sonata for Two Pianos and Percussion', *New Hungarian Quarterly*, xi/40 (1970), 49

R. Travis: 'Tonal Coherence in the First Movement of Bartók's Fourth String Quartet', *Music Forum*, ii (1970), 298

V. Lampert: 'Vázlat Bartók II. vonósnégyesének utolsó tételéhez' [Sketch for the last movement of Bartók's Second Quartet], *Magyar zene*, xiii (1972), 252

I. Rácz: 'Bartók Béla csíkmegyei pentaton gyűjtése 1907-ben' [Bartók's collection of pentatonic tunes in Csík county, 1907], *Népzene és zenetörténet*, i (Budapest, 1972), 9

S. Veress: 'Béla Bartóks 44 Duos für zwei Violinen', *Erich Doflein: Festschrift zum 70. Geburtstag* (Mainz, 1972), 31

V. Lampert, ed.: *Jemnitz Sándor válogatott zenekritikái* [Jemnitz's selected criticism, 1924–1950] (Budapest, 1973)

L. Somfai: 'Bartók rubato játékstílusáról' [Rubato style in Bartók's own interpretation], *Magyar zenetörténeti tanulmányok* (1973), 225

M. Papp: 'Bartók hegedűrapszódiái és a román népi hegedűs játékmód hatása Bartók műveire' [Bartók's violin rhapsodies and the influence of the Romanian peasant violin style on Bartók's works], *Magyar zene*, xiv (1973), 299

T. Tallián: 'Bartók levélváltása R. St. Hoffmann-nal' [Bartók's correspondence with Hoffmann], *Magyar zene*, xiv (1973), 134–85

M. Carner: 'Béla Bartók', *NOHM*, x (1974), 274

F. László: 'Megjegyzések a Cantata profana szövegéhez', 'Még egyszer a Cantata profana szövegéről' [Notes on the text of *Cantata profana*], *Művelődés*, xxviii/9, 11 (1975), 49, 25

D. Dalton: 'The Genesis of Bartók's Viola Concerto', *ML*, lvii (1976), 117

B. Suchoff: 'Bartók in America', *MT*, cxvii (1976), 123

T. Tallián: 'Der Briefwechsel Bartóks mit R.St.Hoffman', *SM*, xviii (1976), 339

J. Breuer: 'Kolinda-ritmika Bartók zenéjében' [*Colinda* rhythms in Bartók's music], *Zeneelmélet, stíluselemzés* (Budapest, 1977), 84

G. Perle: 'The String Quartets of Béla Bartók', *A Musical Offering: Essays in Honor of Martin Bernstein* (New York, 1977)

L. Somfai: 'Bartók's Writings', *MT*, cxviii (1977), 395 [review of *Béla Bartók Essays*, ed. B. Suchoff (1976)]

Bartók

———: 'Strategies of Variation in the Second Movement of Bartók's Violin Concerto 1937/8', *SM*, xix (1977), 161

———: 'Vierzehn Bartók-Schriften aus den Jahren 1920/21: Aufsätze über die zeitgenössische Musik und Konzertberichte aus Budapest', *Documenta Bartókiana*, v (1977), 15

A. Szentkirályi: 'Some Aspects of Béla Bartók's Compositional Techniques', *SM*, xx (1978), 157

A. Wilheim: 'Bartók találkozása Debussy művészetével' [Bartók's encounter with Debussy's art], *Zenetudományi dolgozatok 1978* (Budapest, 1978), 107

T. and P. J. Bachmann: 'An Analysis of Béla Bartók's Music through Fibonaccian Numbers and the Golden Mean', *MQ*, lxv (1979), 72

S. Thyne: 'Bartók's Mikrokosmos: a Reexamination', *Piano Quarterly*, no.107 (1979), 43

I. Oramo: 'Modale Symmetrie bei Bartók', *Mf*, xxxiii (1980), 450

E. Antokoletz: 'The Musical Language of Bartók's 14 Bagatelles for Piano', *Tempo*, no.137 (1981), 8

F. Bónis: 'Bartók und Wagner', *ÖMz*, xxxvi (1981), 134

S. Kovács: 'Reexamining the Bartók/Serly Viola Concerto', *SM*, xxiii (1981), 295

G. Kroó: 'Data on the Genesis of Duke Bluebeard's Castle', *SM*, xxiii (1981), 79

Y. Lenoir: 'Contributions à l'étude de la Sonate pour violon solo de Béla Bartók', *SM*, xxiii (1981), 209

L. Somfai: 'Die "Allegro barbaro": Aufnahme von Bartók text-kritisch bewertet', *Documenta Bartókiana*, vi (1981), 259

———: 'Manuscript versus Urtext: the Primary Sources of Bartók's Work', *SM*, xxiii (1981), 17

———: 'The Rondo-like Sonata Form Exposition in the First Movement of the Piano Concerto no.2', *New Hungarian Quarterly*, xxii/84 (1981), 86

B. Suchoff: introduction to *Piano Music of Béla Bartók: the Archive Edition* (New York, 1981)

T. Tallián: 'Die Cantata profana: ein "Mythos des Übergangs" ', *SM*, xxiii (1981), 135

J. Ujfalussy: 'Béla Bartók: Werk und Biographie', *SM*, xxiii (1981), 5

A. Wilheim: 'Bartók's Exercises in Composition', *SM*, xxiii (1981), 67

S. Kovács: 'Über die Vorbereitung der Publikation von Bartóks grosser ungarischer Volksliedausgabe', *SM*, xxiv (1982), 133

L. Dobszay: 'Absorption of Folksong in Bartók's Composition', *SM*, xxiv (1982), 303

I. Oramo: 'Die notierte, die wahrgenommene und die gedachte Struktur bei Bartók', *SM*, xxiv (1982), 439

100

Bibliography

I. Waldbauer: 'Intellectual Construct and Tonal Direction in Bartók's "Divided Arpeggios" ', *SM*, xxiv (1982), 527

H. Winking: 'Klanflächen bei Bartók bis 1911', *SM*, xxiv (1982), 549

L. Somfai: 'The Budapest Bartók Archives', *FAM*, xxix (1982), 59

M. Gillies: 'Bartók's Notation: Tonality and Modality', *Tempo*, no.145 (1983), 4

——: 'Bartók's Last Works: a Theory of Tonality and Modality', *Musicology*, vii (1982), 120

R. Howat: 'Bartók, Lendvai and the Principles of Proportional Analysis', *Music Analysis*, ii/1 (1983), 69

IGOR STRAVINSKY

Eric Walter White

Jeremy Noble

CHAPTER ONE

1882–1910

I Life

Igor Fyodorovich Stravinsky was born in Oranienbaum
(now Lomonosov) on 17 June 1882. His father, Fyodor
Ignat'yevich Stravinsky, was a fine bass and was engaged
as an opera singer first at Kiev, where he met Anna
Kholodovsky, who became his wife, and then at the
imperial opera house, the Mariinsky, St Petersburg. Igor
was the third of four children, all of them boys. He was
fond of his youngest brother Gury, who became a
promising singer and was killed on the Romanian front
in 1917; but he does not seem to have got on particularly well with the other members of his family. His
youth was divided between winters in St Petersburg,
where his parents had an apartment on the Krukov
Canal, and summers in the country, where the family
visited various estates belonging mainly to his mother's
sisters and their husbands. There was Pavlovka in the
government of Samara, where the Ielachich family lived;
Pechisky near Proskurov and Yarmolintsi, which
belonged to aunt Katerina; and an estate at Ustilug in
Volin which had been bought in the 1890s by Gabriel
Nossenko and his wife Maria.

Stravinsky attended the St Petersburg School no.27
from the age of 11 until he was 15, and went on to the
Gurevich School. He then spent eight terms at St
Petersburg University reading law; but by his own ac-

10. *Stravinsky (extreme left) and his wife Katerina (extreme right) at the home of Rimsky-Korsakov (seated next to Stravinsky) in 1908; also in the picture are Rimsky-Korsakov's daughter Nadezhda, and her fiancé Maximilian Shteynberg*

count he was a bad student at both school and university. It was a different matter with music. At home he frequently heard his father practising his operatic roles; and from an early age he was encouraged to attend ballet and opera performances at the Mariinsky. In later life one of his treasured memories was of seeing Tchaikovsky during the interval of a gala performance to celebrate the 50th anniversary of Glinka's *Ruslan and Lyudmila* a few weeks before that composer's sudden death in 1893.

At the age of nine Stravinsky began to take piano lessons from a Mlle A. P. Snetkova, and she was succeeded a few years later by Mlle L. A. Kashperova, a pupil of Anton Rubinstein. His parents seem to have hoped that in the course of time he might become a professional pianist. He received harmony lessons from Fyodor Akimenko and, later on, instruction in both harmony and counterpoint from Vasily Kalafati. In his teens he found much enjoyment in improvisation, and his interest began to turn to composition. One of his student friends was Vladimir Rimsky-Korsakov, the youngest son of the composer; and when in the summer of 1902 Stravinsky accompanied his parents to Bad Wildungen and discovered that the Rimsky-Korsakov family was in the neighbourhood, he accepted an invitation to visit them, and in the course of this visit found an opportunity to consult Rimsky-Korsakov about his future career. At first he was disappointed by Rimsky-Korsakov's lack of enthusiasm for the little piano pieces he had brought with him; but the older man's counsel was sound. He advised Stravinsky not to enter the conservatory, but to continue his private studies in harmony and counterpoint. He thought his work of composition should be systematically supervised and added that he personally would be prepared to offer advice if consulted. Shortly after this meeting, Stravinsky's father died, on 3 December 1902, and during the next six years Rimsky-Korsakov was to become something of a father figure to the young composer as well as his intimate musical adviser.

Stravinsky's circle of friends and his musical interests now began to widen. He regularly attended the weekly musical gatherings held at the Rimsky-Korsakov house, and some of his friends established the 'evenings of contemporary music', where music by contemporary German and French composers was performed as well as works by young Russians. Here his early Piano Sonata in F♯ minor (see fig.11) received its first public performance. This sonata had caused him some trouble when he started it in 1903; and following Rimsky-Korsakov's advice, he went to consult him in his summer retreat at Lzy. There followed a period of nearly three years during which he received regular instruction from Rimsky-Korsakov, mainly in instrumentation.

In 1905 Stravinsky's university career came to an end, and that autumn his engagement to his cousin Katerina Nossenko was announced. As they were first cousins and there was an imperial statute forbidding such marriages, the wedding had to be a quiet one. It took place in the village of Novaya Derevnya near St Petersburg on 23 January 1906. After a honeymoon at Imatra, the newly married couple settled down in the Stravinsky family apartment on the Krukov Canal for about a year before moving to an apartment of their own on the English Prospekt. The summers continued to be spent in Ustilug, which Stravinsky found congenial to composition; and he now built his own house there on a site provided by his father-in-law. A son, Fyodor, was born to the couple in 1907 and a daughter, Ludmila, in 1908.

After his marriage Stravinsky continued to see Rimsky-Korsakov regularly, showing him his compositions in draft and discussing them with him movement by movement. His first work with an opus number, the Symphony in E♭, was dedicated to Rimsky-Korsakov, who arranged for a private performance to be given by the court orchestra under Wahrlich on 9 May 1907. Later that year Stravinsky showed him his *Scherzo fantastique* for orchestra and the sketches for the first act of his projected opera *Solovey* ('The Nightingale'), the libretto of which he had written in collaboration with his friend Stepan Mitusov. The following winter Rimsky-Korsakov's health began to fail. In spring 1908 Stravinsky told him of his intention to compose an orchestral fantasy, *Feu d'artifice* ('Fireworks'), to celebrate the forthcoming wedding of Rimsky-Korsakov's daughter Nadezhda to Maximilian Shteynberg. This was written at Ustilug; but by the time it was finished Rimsky-Korsakov had died at Lzy. After the funeral Stravinsky returned to Ustilug and composed a funeral dirge in his master's memory. Unfortunately this work, performed that autumn at a Belyayev concert at St Petersburg, was not published, and the musical material disappeared some years later; but Stravinsky remembered it with affection as the best of his compositions before *Zhar'-ptitsa* ('The Firebird').

When the *Scherzo fantastique* and *Fireworks* were performed in St Petersburg they made a deep impression on some members of the audience, including par-

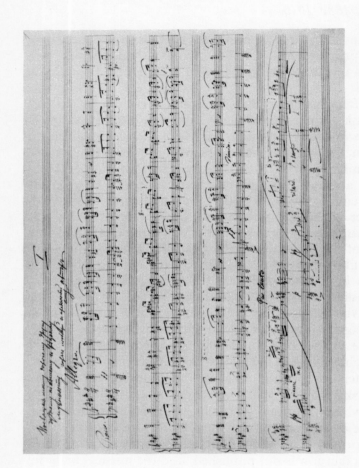

11. Autograph MS of the beginning of Stravinsky's Piano Sonata in F♯ minor, composed 1903–4

ticularly the impresario Sergey Dyagilev, who had been closely associated with special manifestations of Russian art in Paris. For 1909 he was planning a mixed season of opera and ballet; and among his advisers were the dancer and choreographer Mikhail Fokin and two artists of distinction, Leon Bakst and Alexandre Benois, but as yet no musician. After hearing a concert performance of these two works Dyagilev realized that Stravinsky was almost certainly the composer he needed to complete his advisory group. For his 1909 season in Paris he asked Stravinsky to orchestrate Grieg's *Kobold* for a ballet, *Le festin*, and two Chopin piano pieces for *Les sylphides*. For his 1910 season he invited him to write the music for a ballet to be based on the Russian fairy tale of the Firebird after Lyadov had failed to produce a manuscript. This was Stravinsky's first large-scale commission. He began the score when staying at the Rimsky-Korsakov country house at Lzy in the autumn of 1909 and finished it in St Petersburg early the following spring. The work was composed in collaboration with Fokin; the two men brought their sketches and initial ideas to one another and worked out many of the details together. The first performance of *The Firebird* was given by the Ballets Russes at the Paris Opéra on 25 June 1910.

II Works

The Firebird is the earliest of Stravinsky's scores to have won and retained a firm place in the repertory; and yet he was in his 28th year by the time he composed it. It cannot be regarded as a fully mature work, although it is a better and more original one than he himself was later to rate it. This gives some indication of the difficulty that Stravinsky had from the first in finding and forming his individual voice. Style was to be at once a problem and a creative stimulus to him at various periods throughout his long career – and not only because of the vicissitudes of his outward life, with its two exiles, unsettling though these undoubtedly were. For Stravinsky, as for many Russian composers before and since, history posed certain choices of outlook and allegiance, and in his case these were compounded rather than solved by a restlessly inquiring mind and an insatiable appetite for the sonorous raw materials of his art.

The music Stravinsky composed before *The Firebird* is mainly of interest for the way in which it reveals, as clearly as fossils in a succession of geological strata, the influences that came the way of a young man in a cultivated bourgeois home in St Petersburg in the first decade of the 20th century, and the way in which this particular one assimilated them. Of the two posthumously published early piano works the little Scherzo of 1902 is a very modest genre piece, some way after Tchaikovsky, whose only pointer towards things to come is the displaced chordal accent in its third bar; but the Sonata of 1903–4, the first fruits of his studies with Rimsky-Korsakov, is altogether more ambitious. Here again Tchaikovsky is the clearest influence – most obviously in the scherzo, but also in the more weighty first and last movements; in these Tchaikovsky's own father figure, Schumann, can be detected both in the textures and in the characteristic dotted rhythms. This is the work of a would-be cosmopolitan Petersburg composer for whom 'sonata' inevitably implied an adherence to German tradition, though there is little or nothing of the Beethoven influence that Stravinsky himself, in much later years, thought he remembered. Clearly he had already, in his early 20s, acquired a sound if conventional grasp of 'piano style'; what is weak in the sonata is not so much the disparity between its various ingredients as the clumsiness of the transitions between them.

This is a weakness, of course, only in a composer who accepts certain traditional formal assumptions, and Stravinsky was to discover himself precisely by jettisoning these. But not at once. For the present he was still bent on conforming to them, and the Symphony in E♭ op.1, composed under Rimsky-Korsakov's close supervision, shows a clear gain in professionalism of this conventional kind. Along with the increased technical assurance there goes a new (compared with the sonata) emphasis on folk-influenced melodic material, and this gives the music much more rhythmic variety and vitality. Tchaikovsky reappears in the scherzo (though the trio melody seems more like a domesticated relation of the *Petrushka* nursemaids), but the strongest influence apart from that of Rimsky-Korsakov now seems to be that of Borodin, and no doubt there is some debt to Glazunov too, though Stravinsky's personal dislike of the man may later have led him to play this down. The young composer shows his individuality in his occasional tentative forays into Wagnerian orchestral effect (e.g. the tremolandos at fig.24 of the first movement) and chromatic harmony. The stylistic inconsistency which this produces was later to be turned to valid dramatic use in *The Firebird*, and the end of the rather Franckian slow movement again foreshadows that score (the transition from the 'Berceuse' to the final scene) – but such indications of the future are, perhaps naturally, more pronounced in the smaller-scale, less 'symphonic' works of the period.

In *Favn' i pastushka* ('Faun and Shepherdess'), a setting for mezzo-soprano and small orchestra of words by Pushkin composed in 1906, the appearance of the Faun at the beginning of the second song is accompanied by an orchestral gesture that immediately recalls the appearance of the Firebird herself, and the very end of the song cycle clearly prefigures that of the ballet, even as regards key. Tchaikovsky is still present at times, particularly in the yearning phrases of the first song, but with a stronger admixture of Wagner; Rimsky-Korsakov was worried by a tincture of Debussian modernism that was more conspicuous to him than it is to a modern listener. The vocal writing is rhythmically rather unadventurous, and scarcely less so in the two Gorodetzky songs for mezzo and piano composed during the two following summers – but in this Stravinsky may already have been reflecting the syllabic character of Russian folksong. What is more interesting is the curiously empirical chromaticism of the harmony; progressions are not marked by any strongly marked sense of a controlling bass, but rather edge from one chord to another, as if worked out at the piano. This is also a characteristic of the Four Studies for piano, composed in the summer of 1908, but in them it is exploited more systematically – and the first two, at least, are also systematic in their exploitation of opposed rhythms in the two hands (3 or 2 against 5; 6 against 4 or 5), which give the music a peculiarly Skryabinesque fluidity. In the third study Stravinsky is for once closer to Rakhmaninov, but in the last, and best, he is already clearly himself; the sense of harmonic drive in this *moto perpetuo* is unmistakably original.

The distance Stravinsky had already travelled in the four or five years since the sonata is remarkable, but no doubt the Studies seemed less important at the time, to him and his audiences, than the two short pieces for large orchestra of 1907–8, the *Scherzo fantastique* and the fantasy *Fireworks*. Both of these show, far more clearly than the songs or the piano music, the impact (so much deplored by Rimsky-Korsakov) of the new French music. Dukas' *L'apprenti sorcier* makes an almost literal appearance in the middle section of *Fireworks*, but the most pervasive influence is that of Debussy's *Nocturnes*, which Stravinsky had heard at one of Ziloti's concerts in St Petersburg; he himself would later describe it as 'among the major events of my early years'. But the most major event was unquestionably his being commissioned by Dyagilev, on the strength of these two scores, to compose a ballet for his company's 1910 season in Paris. Here for the first time Stravinsky was faced with the challenge of composing an orchestral work nearly an hour in length, and of telling a dramatic story through it with no irksome formal constrictions. For the first time, too, he was stimulated to explore what turned out to be one of his most characteristic gifts, the ability to express physical gestures and movements (and the psychological states that prompt them) in purely musical terms – a gift in which he has had no rival since Wagner.

In retrospect Stravinsky claimed to have been already out of sympathy with the aesthetic premises of the folk-tale scenario, above all with its traditional division into set dances and mimed action, reflecting the arias and recitatives of opera, but if this was so he concealed the fact with remarkable success: the dialogue between Ivan and Kashchey, for instance, which he criticized for its over-literalness, is incomparably vivid and assured. It is hard not to feel that Stravinsky may have read back into his attitude at this time the strongly anti-'realist' stance he was to take up later (for an account of this specifically Russian aesthetic controversy see R. Taruskin: 'Realism as Preached and Practised: the Russian *opéra dialogué*', *MQ*, lvi (1970), 431). In later years he was concerned mainly to praise the score's technical novelties of orchestration – notably the ingenious natural-harmonic arpeggios at bar 14 – and it is true that it was clearly designed to outdo Rimsky-Korsakov in sheer picturesqueness. But Stravinsky also borrowed from his late master the device (used in *The Golden Cockerel*) of using chromaticism and a melodic and harmonic vocabulary based on the augmented 4th for the supernatural characters, good and evil, and a modal-diatonic style for the human protagonist, Ivan, and the enchanted princesses whom he rescues. This does much to pull together the music's centrifugal repertory of styles. In addition to the influences already mentioned, the Firebird's dance of supplication is pure Balakirev, and Ravel's *Rapsodie espagnole* is laid under contribution in her 'Berceuse'; but although it can be (and has been) objected that *The Firebird* is a thesaurus of current styles, this in no way weakens its impact in the theatre, where every section seems apt to its own place in the action. Stravinsky himself later expressed some doubts about the 'Mendelssohnian–Tchaikovskian' idiom of the Princesses' scherzo, yet in context this seems as completely in keeping as the fantastic music invented for the demon Kashchey and his grotesque retinue. Whatever reservations the sophisticated young composer may have felt about the piece as he was composing it, the musical vitality which has kept the concert suites, at least, in the repertory is some indication that he was genuinely stimulated by the project as a whole.

CHAPTER TWO

1910–14

I Life

The success achieved by *The Firebird* altered the course of Stravinsky's life. At that time Paris was the international centre of the world of art; the Ballets Russes one of its prime sensations; and Stravinsky's the most important original score in the ballet repertory. This meant that overnight he became known as the most gifted of the younger generation of Russian composers, and during the next few years his music became better known and appreciated in western Europe than in his native Russia. Before the end of the Ballets Russes season in Paris that summer, he brought his wife and family over from Ustilug to share his triumph with him. It was clear that his musical future would be closely bound up with the fortunes of the Ballets Russes, and as the company looked on Paris as its base, he would have to be prepared to spend a good part of his time in western Europe.

Fortunately there was no dearth of ideas for new works. Already in April or May 1910, when finishing the full score of *The Firebird* in St Petersburg, he had had a dream which gave him the idea of writing a symphonic work based on a pagan ritual sacrifice. In Paris that summer he mentioned this to Dyagilev, who encouraged him to proceed; but when Dyagilev visited Stravinsky later that summer in Lausanne, where

Katerina was expecting the birth of her third child (Svyatoslav Sulima), he found to his surprise that the composer had started to write a completely different work, a kind of Konzertstück for piano and orchestra. As Stravinsky had had in mind a 'picture of a puppet, suddenly endowed with life, exasperating the patience of the orchestra with diabolical cascades of arpeggios', he had provisionally entitled the piece *Petrushka*. Dyagilev immediately saw the dramatic possibilities of the subject and managed to persuade the composer to alter the course of the work and turn it into a ballet score. Benois was chosen to be his collaborator; and *Petrushka* was composed that autumn and winter in Lausanne, Clarens and Beaulieu, with a break in the composition at Christmas when Stravinsky returned to St Petersburg for a few days to attend meetings with Dyagilev, Fokin, Nizhinsky and Benois about the nature of the ballet.

Petrushka was first performed at the Théâtre du Châtelet, Paris, on 13 June 1911 and proved just as successful with the public and critics as *The Firebird* had been; but it was undoubtedly a more original work. In the first place, Stravinsky had been able to play a leading part in the construction of the scenario, which had not been the case with *The Firebird*. Secondly, whereas the music of *The Firebird* showed that the pupil had learnt all that his master had had to teach him, in *Petrushka* for the first time the authentic voice of the new master is heard.

After the Ballets Russes Paris season that summer, Stravinsky retired to Ustilug where he remained working on *Vesna svyashchennaya* ('The Rite of Spring'), as the 'great sacrifice' project was now called, until the autumn, when he and his family moved to Clarens,

12. Stravinsky
(immediately
behind seated
lady) with
members of the
Ballets Russes at
Monte Carlo,
1911; among
those present are
Dyagilev, Benois,
Nizhinsky and
Karsavina

120

Switzerland. By the New Year the first half of the score was virtually complete; but now it became apparent to Dyagilev that he would be unable to mount *The Rite of Spring* in the summer of 1912 as originally planned. So its production was postponed a year; and this meant Stravinsky could work more slowly on the remainder of the score. It also enabled him to complete *Zvezdolikiy* ('The King of the Stars'), a short cantata for male choir and large orchestra setting a transcendental poem by Bal'mont, before leaving Ustilug to attend the Ballets Russes seasons in Paris and London that summer (1912). By this time he was a well-known figure in Parisian musical circles and on friendly terms with Debussy, Ravel and numerous other musicians and celebrities. After his visit to London (his first) he returned to Ustilug for the remainder of the summer – a stay that was interrupted by an invitation from Dyagilev to join him in Bayreuth for a performance of *Parsifal* – and in the autumn the Stravinsky family moved back to Clarens. By then the score of *The Rite of Spring* was virtually complete, and Stravinsky had started to set three Japanese lyrics for soprano and small chamber ensemble (or piano). This left him free to visit Berlin, Budapest, Vienna and London during an extended Ballets Russes tour between November 1912 and February 1913. It was in the course of this Berlin visit that he met Schoenberg for the first time and heard a performance of his *Pierrot lunaire*, which impressed him deeply. After London he returned to Clarens, where Ravel joined him, and they collaborated on an adaptation of Musorgsky's *Khovanshchina* for production by the Ballets Russes in Paris that spring.

The first night of *The Rite of Spring* (Théâtre des

13. Stravinsky in his study at Ustilug, 1912

Champs-Elysées, Paris, 29 May 1913) gave rise to one of the great theatrical scandals of all time. Even during the orchestral introduction mild protests against the music could be heard. When the curtain rose the audience became exacerbated by Nizhinsky's choreography as well as Stravinsky's music, and protests and counter-protests multiplied. At times the hubbub was so loud that the dancers could not hear the music they were supposed to be dancing to. Audiences at subsequent performances in Paris and London that summer behaved with normal decorum; but to those present on the first night the riot in the theatre was a traumatic experience. As for Stravinsky, a few days after the first night he fell ill with typhoid fever and had to spend several weeks in a nursing-home at Neuilly. When he recovered, he returned to Ustilug for the rest of the summer and was back in Clarens with his family in the early autumn. He had recently received a commission from the newly founded Free Theatre of Moscow to complete his opera *The Nightingale* for production there in 1914. Act 1 had been written at Ustilug in 1908–9. Acts 2 and 3 were now added at Clarens and Leysin where the family moved early in 1914, because Katerina Stravinsky, who was expecting her fourth child (a daughter, Milena), had fallen ill with tuberculosis and needed hospital treatment. In the event the Moscow Free Theatre project collapsed; and Dyagilev offered to assume responsibility for the production of *The Nightingale* by the Ballets Russes during its seasons in Paris and London that summer – an offer that was accepted.

The Stravinsky family did not return to Ustilug that summer but remained in Switzerland, moving from

Leysin to Salvan, where Stravinsky composed three pieces for string quartet. He was now contemplating the idea of a cantata celebrating Russian village wedding customs. Realizing that his library in Russia contained some useful works dealing with Russian popular verse and song that he would need for this cantata (subsequently entitled *Svadebka*, 'The Wedding'), he made a hurried trip to Ustilug and Kiev in mid-July; shortly after his return to Switzerland war broke out.

II Works

It may be some indication of the primary importance that Stravinsky was coming to attach to formal coherence that he should originally have conceived that most dramatic of ballets, *Petrushka*, as a concert piece for piano and orchestra. The 'advanced' idiom characteristic of Petrushka's own music (in the ballet's second scene) is thus the heart of the work; the 'normal', extrovert music of the carnival scenes, with their post-Rimskian exuberance of rhythm and colour, and their strongly folk-influenced melodic vocabulary (see F. W. Sternfeld: 'Some Russian Folksongs in Stravinsky's *Petrouchka*', *Notes*, ii (1944–5), 95) thus represents the relatively conventional shell from which the chicken of Stravinsky's own style was beginning to emerge. These crowd scenes are characterized by the use of largely diatonic harmony, often with extended ostinatos (notably the accordion-like alternation of 5ths and 3rds) and internal pedals. A basic regularity of rhythmic pulse is cunningly varied by episodes of calculated asymmetry and syncopation. The 'Russian Dance' which forms the last part of the first tableau, with its hammered parallel triads and diatonic added-note ostinatos, clearly has much in common with this idiom, but the other section deriving from the original concert-piece kernel of the work, 'Petrushka's Cry' as it was originally called, is very different. Here the diatonic language alternates with, and is sometimes combined with, a far more dissonant harmonic idiom based essentially on bitonal combinations, of which the famous Petrushka chord (C major and F♯ major superimposed) is the most important. Yet it should be noted that even in these sections, for all their elaboration of harmony, instrumental colour and rhythm, the melodic language itself is still essentially diatonic.

This essential dichotomy between melody and harmony is still to some extent a feature of the third of Stravinsky's famous pre-war ballet scores, *The Rite of Spring*, but here burlesque and parodistic elements have been shed completely, and the result is a far greater degree of homogeneity than in *Petrushka*. This is already foreshadowed in the two pairs of songs composed in 1910 and 1911. The Verlaine settings, in keeping with their texts, are very French in their musical language too: whole-tone phrases and an occasional chord of the 13th suggest Debussy and Ravel respectively. But a passage of stepwise descending minor chords against a rising sequence of 7ths in the bass (on the word 'apaisement') points unmistakably to the prelude to the second part of *The Rite of Spring*; likewise in the second of the much barer and more linear Bal'mont songs Stravinsky experimented briefly with an ostinato of superimposed duple and triple rhythms which he was later to take up in more extended fashion in the introduction to the whole ballet. As chordally conceived as the Bal'mont songs are linear is the short cantata on a text by the same poet, for male-voice choir and huge orchestra, *The King of the Stars*. Dedicated to Debussy, this was Stravinsky's most single-minded exploration to date of advanced harmony. Its bitonality frequently results in characteristic simultaneous major–minor formations, such as had already appeared momentarily in *Petrushka* and were soon to be used much more frequently in *The Rite of Spring*. Probably because of the mystical nature of the text, however, Stravinsky here affected a quasi-liturgical uniformity of rhythm, and the music makes a somewhat stagnant impression when compared with, for example, the 'Mysterious Circles' section of the latter ballet.

A connection between the Bal'mont songs and *The Rite of Spring* has already been pointed out; it is a curious but convincing indication of the organic nature of Stravinsky's development that a work is often foreshadowed in a brief passage in one of its predecessors. In the same way, towards the end of the gestation of one section of a work, his imagination would often throw up a forerunner of the following one, as the published sketches of *The Rite of Spring* demonstrate remarkably clearly (see fig.14). They also show that Stravinsky's initial ideas, however abstractly he might later choose to develop them, were at this stage almost always generated by specific visual images, and that they fall essentially into two categories, melodic and harmonic. The harmonic 'germ' of *The Rite of Spring* is again a bitonal combination of two adjacent chords – this time of E♭ major (with added minor 7th) and F♭ major; it is presented both as a running semiquaver *moto perpetuo* and congealed into a repeated chord. Melodic ideas, on the other hand, are often surprisingly straightforward in their in-

14. First page of Stravinsky's sketches for 'The Rite of Spring', 1911–13

itial form, but are then subjected to radical transform-
ations, above all of note order, tempo and rhythm.
Rhythm, in fact the most strikingly original aspect of
the score, is not a purely spontaneous manifestation, but is
the product of quite as much intense creative effort as the
harmony. Yet it would be wrong to regard the rhythmic
element as being in any sense arbitrary or unrelated to
the rest: on the contrary, the convulsive asymmetry of
The Rite's rhythms should rather be seen as a direct
function of its harmonic tensions. This score, whose
subject matter concerning the ambiguous triumph and
cruelty of spring and the process of natural renewal
evidently stirred Stravinsky to his creative depths,
mark both a high level of inspiration and an extreme
point along one line of technical development.
Stravinsky had found not only a voice, but an utterly
individual vocabulary, yet it was a vocabulary that
could essentially be used only once. The quest for a
style, conceived not as a response to particular expressive needs but as a language for civilized everyday
discourse, had to continue.

On this quest *The Nightingale* represents something
of a side-road. The first act, having been composed even
before *The Firebird*, is stylistically still very much
indebted to Debussy (it starts with an almost literal
quotation from the *Nocturnes*, this time 'Nuages'); both
the vocal lines and the harmony are closely related to
those of the Verlaine songs. But by the time Stravinsky
took the score up again in the autumn of 1913 he had
lived through four years of the most rapid development
that any composer has ever experienced, a coming-of-
age all the more intense for being belated; it was impossible for him to return to the initial style. Fortunately,

however, the story's own in-built contrast between rustic innocence and the corruption of the imperial court provides at least a superficial justification for the marked change in style. Stravinsky evidently enjoyed concocting the harmonically and instrumentally sophisticated chinoiserie of the court scenes, though their artificiality inevitably seems a little contrived after the blazing urgency of *The Rite of Spring*. The most original music is probably that of the third act; here, as in the slightly earlier *Tri stikhotvoreniya iz yaponskoy liriki* ('Three Japanese lyrics'), the violence and imaginative abundance of the ballet has been distilled into music of the most extreme refinement.

CHAPTER THREE

1914–20

I Life

The advent of war inevitably dislocated the pattern of Stravinsky's life. Whereas during the past four years the family visits to Switzerland had been occasional trips undertaken partly for health reasons and partly so that he could be within easy reach of Paris, now Switzerland became a permanent haven of refuge. Wartime conditions made it impossible for the Ballets Russes to function on normal lines in western Europe; Stravinsky's music publishers were mostly in enemy territory (Germany); and he could no longer depend on the income from his Russian estate. He began to find himself in financial straits; it was only as the result of a generous gift of money from Thomas Beecham that he was able to finance his mother's return to St Petersburg at the outbreak of war.

The composition of *The Wedding* occupied a considerable amount of his time, though he no longer had a deadline to work to. First he constructed the libretto, a selection of words, phrases, sentences from Kireyevsky's collection of folksongs. Then he set the words for four-part chorus and four soloists. There were numerous interruptions to the work; but the short score was completed on 11 October 1917. It was dedicated to Dyagilev, who had been moved to tears when he first heard Stravinsky play the sketches for scenes i and ii.

However, it took Stravinsky another six years before he found the right instrumental formula for the accompaniment to the voices. The instrumentation was eventually finished on 6 April 1923 at Monaco; and the Ballets Russes gave the first performance at the Théâtre de la Gaîté Lyrique, Paris, on 13 June 1923.

Stravinsky's intensive researches into Russian folk material at the time he was working on *The Wedding* produced a number of by-products – the text for a farmyard burlesque, *Bayka* ('Reynard'), and various groups of Russian songs, particularly the *Pribaoutki* and the *Berceuses du chat* ('Cat's Cradle Songs').

Shortly after the outbreak of war the Stravinsky family rented a house in Clarens belonging to the young Swiss conductor Ernest Ansermet for part of the winter; and in 1915 they moved to Morges, where they stayed until 1920. As Switzerland was neutral Stravinsky was able to undertake a few trips abroad. He spent a fortnight with Dyagilev in Florence in autumn 1914. The following February he visited Dyagilev in Rome, where they both attended the first Italian concert performance of *Petrushka* (conducted by Casella). On this occasion he met Gerald Tyrwhitt, later Lord Berners, and Prokofiev, newly arrived from Russia. The following month Dyagilev visited Switzerland, staying at Ouchy, where he assembled a small group of artists in view of the forthcoming visit by the Ballets Russes to the Metropolitan Opera House, New York. Ansermet was to be in charge of the orchestra, and before leaving for the USA he effected two introductions that were to prove of importance to Stravinsky during this period.

First he introduced him to C. F. Ramuz, a well-known author in French-speaking Switzerland. The acquaintance with Ramuz blossomed into friendship; and the Swiss writer supplied Stravinsky with French versions of the texts of *The Wedding*, *Reynard* and various groups of songs, although he knew no Russian and had to rely on Stravinsky's literal translations for his basic material. In 1918 the two collaborated over the text of *Histoire du soldat* ('The Soldier's Tale'), which (on Stravinsky's suggestion) was based on one of the stories in Alexander Afanas'yev's collection. Later, in 1929, Ramuz published an account of their friendship, *Souvenirs sur Igor Strawinsky*, which, though warm in tone and perceptively written, did not please the composer, who felt his privacy had been unjustifiably invaded. Ansermet's second introduction was to Aladar Racz, a Hungarian cimbalom player, whose virtuoso performance at Maxim's, Geneva, inspired Stravinsky to make himself proficient in playing the instrument and to write cimbalom parts in several of his works, including *Reynard* and *Rag-time* (1918).

Just before the Ballets Russes left for the USA Dyagilev arranged two charity matinées for the Red Cross, one at Geneva (20 December 1915) and one in Paris nine days later. Stravinsky was engaged to conduct *The Firebird* in both programmes, the concert suite at Geneva and the ballet in Paris; and these were his first public appearances as a conductor, though at Ansermet's suggestion he had taken the Montreux Kursaal Orchestra through his Symphony in E♭ at a concert rehearsal in April 1914.

After the company set sail for New York, Stravinsky called on his friend the Princess Edmond de Polignac in Paris, and she offered to commission a small chamber work from him. He had already made some preliminary

15. Front cover by Picasso for the first edition (1919) of Stravinsky's piano arrangement of his 'Rag-time'

sketches for *Reynard*; and it was now agreed he should finish this, although there was no immediate likelihood of its production.

The Ballets Russes's American visit finished at the

end of April 1916, and the company then returned to Spain. Stravinsky joined Dyagilev in Madrid for a few weeks in May, and to this visit may be attributed various Spanish elements in some of his works of this period, such as the Study for pianola, the 'Española' (one of the Easy Pieces for piano four hands), and the 'Royal March' in *The Soldier's Tale*. When the Ballets Russes returned to the USA later in the year, Dyagilev remained behind and, in the course of discussing future plans with Stravinsky, suggested *The Nightingale* might be presented in ballet form. Stravinsky countered with the proposal that the homogeneous music of Acts 2 and 3 should be turned into a symphonic poem (without voices). This was done, the resulting work being *Chant du rossignol* ('The Song of the Nightingale'); later it was used as a ballet score.

In spring 1917 Dyagilev went to Rome, where he arranged four performances at the Teatro Costanzi, the first of which was a gala in aid of the Italian Red Cross. The programme included *The Firebird* and *Fireworks*, both of them conducted by the composer, and the latter with special lighting effects by the Italian futurist painter Giacomo Balla. The news of the Russian Revolution of February 1917 had recently come through: so it seemed inappropriate to begin the performance with 'God Save the Tsar', and Stravinsky arranged the *Song of the Volga Boatmen* at short notice for wind and percussion as a substitute.

At first the revolutionary news from Russia seemed encouraging to a liberal like Stravinsky. On 24 May 1917 he telegraphed his mother, who was still living in the Stravinsky family apartment in Petrograd: 'All our thoughts are with you in these unforgettable days of joy for our beloved Russia freed at last'. But by October the position had radically changed; gradually he realized that the Bolshevik Revolution had made it impossible for him to return to his native land, and he and his family would probably have to live their lives in permanent exile.

The beginning of 1918 was a particularly dark moment in the conduct of the war, and this affected Stravinsky and several of his friends. One day he and Ramuz had an idea. Why not write something quite simple, for two or three characters and a handful of instrumentalists, something that could be played in modest conditions, in village halls and the like? Out of this was born the idea of *The Soldier's Tale*, a piece 'to be read, played and danced'. A backer was found (Werner Reinhart of Winterthur); the Lausanne Theatre hired; and on 28 September 1918 the new work had its first performance with Ansermet conducting. This was a great success; but unfortunately a severe outbreak of influenza forced the organizers to cancel the tour that had been planned to follow the Lausanne performance.

When the war ended it was no longer necessary for Stravinsky to continue to live in Switzerland. At first he thought of settling in Rome, but ultimately decided in favour of France. He did not leave Switzerland until summer 1920; and the latter part of 1919 and the early months of 1920 were spent in writing a new ballet score for Dyagilev, who had suggested he might adapt some music by Pergolesi for a ballet with a *commedia dell'arte* theme. The choreography was by Massin; the décor and costumes by Picasso. Stravinsky obviously fell in love with the Pergolesi pieces that had been put at his disposition; and the resultant ballet *Pulcinella* was a

135

great success when produced at the Paris Opéra (15 May 1920).

II Works

The true and logical successor to *The Rite of Spring* in Stravinsky's output is not *The Nightingale* but *The Wedding*. The first idea of this work occurred to him while he was engaged on *The Rite*, but it was not until 1914 that he was able to think about it seriously, not until 1917 that the music was essentially complete, and not until 1923 that it received its definitive instrumental form. *The Wedding* is thus the central work of this entire period. As profoundly concerned with the cycle of regeneration as *The Rite of Spring*, it represents a kind of Orthodox counterpart to the imaginary pagan ritual of that work and, whether consciously or not, a civilized 'social' answer to its explosive individualism. Such a function demanded a very different style, and it is not surprising to find, among the compositions of the early war years, several which can be regarded as studies for the main work in hand.

First in point of time, and most seminal in content, are the Three Pieces for string quartet (1914). In these Stravinsky seems to have isolated, as in a process of self-analysis, three of the basic musical components of his creative personality. The first, a stylized Russian dance, explores the effects of repetition and permutation of a brief melodic cell against a constant but irregularly grouped pulse; the second, apparently suggested by the antics of the English clown Little Tich, substitutes spastic rhythms and extreme contrasts of theme and texture; the third is a frozen litany, whose harmonic tensions have reached a point of almost complete stasis – the

16. *Choreography by Natalya Goncharova for four women in Stravinsky's 'The Wedding', first performed at the Théâtre de la Gaîté Lyrique on 13 June 1923: pen and ink drawing heightened with white ink*

forerunner of many of Stravinsky's later codas. The ideas which these pieces encapsulate were too basic to find their fulfilment immediately, and the first of the works in which Stravinsky explored more specifically the possibilities of a Russian folk melos wedded to an uncompromisingly individual harmonic style is *Pribaoutki*, settings of nonsense-songs in which a tight-knit melodic line is set against an accompaniment containing a harmonic tension that places it in a new and piquant light – as in the first song, where the accompaniment to an E♭ melody concerns itself principally with the notes F♯, G and A♭, or the last, in which a melody with A as a tonal centre is set over a G♯ drone. The traditional tonal hierarchy of dissonance and resolution is being exploited negatively, in the setting-up of tensions which are deliberately not resolved. The result is a certain stiffness of harmonic language, analogous to aspects of cubist painting; tonal centres, ambiguous or not, are established by repetition, and replaced not by modulation but by simple shift. These, in fact, are only technical procedures of *The Rite of Spring* applied on a smaller scale; what is new is the affectionate exploration of a melodic idiom dictated by Russian popular verse. None of these works (unlike *Petrushka*, for instance) uses genuine folk melodies, but all are impregnated with their style, and Stravinsky seized with particular delight on one feature of the original melodies, their metrical independence of the word-stress. This was to become a feature of his own setting of texts in whatever language, French, Latin, English or Hebrew. The *Cat's Cradle Songs* of 1915–16 are even shorter and simpler, with *Pribaoutki*'s accompaniment of mixed wind and strings replaced by the homogeneous timbre of three clarinets (of various sizes), suggested by the 'cat-like' character of the poems.

Animals are much in evidence in these Russian-style works of the early war years – partly, no doubt, because they occur frequently as characters in the popular texts with which Stravinsky was preoccupied, but also because circumstances had cooped him up in Switzerland with a family of four small children; the nursery world of farmyard tales attracted him as a valid stylization, a mask through which human characteristics could be portrayed without the invitation to identification which had come to seem banal. The world of folk poetry also provided a much needed temporary escape from the hothouse sophistication of the Dyagilev–Paris ambience, even if Stravinsky did, in the event, carry his new-found interest back there with him. *Reynard*, at any rate, the most extended essay in this stylized farmyard vein, was designed for a princess's Paris drawing-room. Its larger format entails an increase in forces: only five solo strings (used for the most part percussively), but seven wind and a group of percussion instruments headed by Stravinsky's new discovery, the cimbalom, whose nasal timbre dictates much of the score's character. It is also notable that the four male voices in no way represent the individual characters in the narrative: the action is danced and mimed, while the voices, placed in the orchestra, singly or collectively utter the characters' words at any given moment. Stravinsky's 'anti-realism' at this stage led him to reject completely the implied mimesis of traditional opera, though he would later come to terms with it on occasion.

The Wedding (conceived earlier, though completed later) shares this characteristic with *Reynard*, but in

139

17. Design (1921) by Mikhail Larionov for Stravinsky's 'Reynard', first performed at the Paris Opéra by the Ballet Russe de Monte Carlo on 18 May 1922: watercolour

other ways is rather different. Where *Reynard* is a large-scale small work, consisting of a number of contrasting episodes linked primarily by their instrumentation and by a very few literal recurrences (hence the need for the framing march that brings the players on and escorts them off at the end), *The Wedding* is, in the widest but the truest sense, symphonic. Melodically it is unified by a close cellular kinship between the various themes that emerge from its texture for development and transformation; rhythmically it is perhaps the most closely knit of all Stravinsky's works. Recognizing that tonal relationships were, for him, as for Debussy, no longer capable of sustaining a lengthy symphonic argument, and perhaps recognizing too that the profound instinctive unity of *The Rite of Spring* could not always be relied on, Stravinsky had looked for other methods of ensuring continuity and devised a form of large-scale rhythmic development more pervasive than anything in his earlier music. Every new tempo in the work is proportionally geared to its predecessor, and so ultimately to a basic pulse that can be felt, if not consciously heard, to run through the entire work; this contains its extreme (and steadily increasing) contrasts of mood, which find their ultimate resolution in the ritual greeting of the newly wedded bride by her husband, and the almost liturgical bellstrokes that consummate their union. It is a procedure common to many, perhaps most, of Stravinsky's succeeding works, but nowhere is it more essential to the music's inner life than in this, the one for and through which it was evolved. No doubt it was in order to emphasize the immensely important role of rhythm in the score that Stravinsky, after experimenting with other, more highly coloured possibilities, eventually decided on a 'black and white' instrumentation, using only pianos and percussion. This has probably militated against *The Wedding*'s popularity, but unlike *The Rite of Spring* it cannot be vulgarized into an orchestral tour de force, and retains, both in and out of the theatre, its unique impact.

By the end of the war years Stravinsky was already beginning to react against his own intense preoccupation with Russian folk material. *The Soldier's Tale*, though based on that material, deliberately aims at transcending national limitations. A Russian accent may still be detected in some of the music, but it is subordinated to such new influences as the Spanish pasodoble band (in the 'Royal March'), tango and ragtime – not to mention the chorales which mark the work's dramatic climax. In this sense, just as *Reynard* sums up the Russian studies, *The Soldier's Tale* sums up the parodistic vein which goes back as far as *Petrushka* and which had been explored more recently in the two sets of pieces for piano duet and in the Study for pianola (1917), based on impressions of Madrid night music. The impact of jazz (or rather ragtime; Stravinsky's knowledge of the idiom was so far limited to printed music brought to him from the USA by Ansermet) can also be heard in two further works of the period: *Rag-time*, for a group of 11 instruments including *Reynard*'s cimbalom but closer in composition to that of *The Soldier's Tale*, and the *Piano-rag-music* written for Artur Rubinstein; both pieces aim rather at a cubist impression of the style than at direct parody in the manner of Les Six. Stravinsky had already passed beyond the stage when he could derive much from jazz apart from a passing frisson, and he would not return to it until the *Ebony Concerto* written for Woody Herman in 1945.

Far more important for Stravinsky's future development was the fortuitous involvement with 18th-century music brought about by Dyagilev's invitation to arrange a ballet score from music by (or attributed to) Pergolesi. Although all that Dyagilev had in mind was an arrangement comparable to Tommasini's of Scarlatti in *The Good-humoured Ladies*, Stravinsky's imagination was more deeply stirred than either of them could have foreseen. The act of composing against the background of a harmonic and rhythmic system as regular and familiar as that of Pergolesi, and with a similarly restricted orchestral palette, made Stravinsky aware how powerful an effect could be achieved, within such a context, by quite small and subtle dislocations. For much of its length *Pulcinella* stays too close to the original to be quite fully Stravinsky, though there is scarcely a bar that does not reveal his presence by some felicity of phrasing, harmony or sonority; but the experience of composing it did undoubtedly point him towards the consistent stylistic persona for which he had been consciously or unconsciously searching – one that would, in the event, serve him for some 30 productive years.

CHAPTER FOUR

1920–39

I Life

Stravinsky brought his family from Switzerland to Brittany during the summer of 1920; and in September Gabrielle Chanel placed her home in the Paris suburb of Garches at his disposal. The following year the family moved, first to a house at Anglet near Biarritz, and later to a villa in the centre of Biarritz. Although he continued to compose at home, his musical activities in Paris led to his making fairly regular use of a studio which the Pleyel company put at his disposal in their pianola warehouse. At that moment he was particularly interested in the recording of his works for player piano, just as later on he was to spend much time supervising the recording of his works for gramophone.

The first major work to be finished after Stravinsky left Switzerland was the Symphonies of Wind Instruments. He had contributed a short instrumental chorale to a special musical supplement that the *Revue musicale* issued in memory of Debussy and this became the final section of the Symphonies, the last major work to be written in what had become his characteristic Russian idiom. In 1921 he joined the Ballets Russes while it was on tour. He went to Spain in April when the company was playing in Madrid, and spent Easter with Dyagilev in Seville. In the summer the company visited London and Stravinsky attended the first concert performance in England of *The Rite of Spring*, conducted by Eugene Goossens (Queen's Hall, 7 June). Sergey Koussevitzky, who happened to be in London at the same time, decided at short notice to include the Symphonies of Wind Instruments in a concert he was giving at the Queen's Hall. The work seems to have been under-rehearsed and badly presented (10 June). But, apart from that fiasco, Stravinsky's London visit appears to have been an enjoyable one; and it gave him the chance to hold a number of important talks with Dyagilev. In the first place, Dyagilev told him about his plan to revive *The Sleeping Beauty* at the Alhambra, London, for a run; and Stravinsky agreed to help with the arrangement of the score. Then Stravinsky told Dyagilev of his intention to compose an *opera buffa* on a story by Pushkin (*Mavra*). Boris Kochno, whom the theatrical designer Serge Soudeikine had recently introduced to Dyagilev, and whom Dyagilev had engaged as his personal secretary, was chosen as librettist.

Earlier that year Stravinsky had met Soudeikine's wife, Vera (née de Bosset), who was to appear in the non-dancing role of the Queen in the revival of *The Sleeping Beauty*. He fell deeply in love with her, and she with him. During the next 18 years they saw as much of each other as possible – mainly in Paris, but sometimes also when she was able to accompany him on concert tours. One of the first results of this liaison was that his next composition after *Mavra*, the Octet for wind instruments (1922–3), was dedicated to her, though the dedication did not appear in the published edition of the work.

Shortly after the Russian Revolution Stravinsky had

written to Arthur Lourié, then commissar of music in Petrograd, asking him to help his mother obtain a visa to leave Russia for France. Permission for her to emigrate was held up for some years; and by the time it came through Lourié himself was in Paris, where he was introduced to Stravinsky by Vera Soudeikine and became his musical assistant for several years. Ultimately, Stravinsky's mother was allowed to travel by boat through the Baltic to Stettin, where she was met by her son in the summer of 1922. She joined the family in Biarritz and accompanied them on their later moves, to Nice in 1924, to Voreppe near Grenoble in 1930 and to Paris in 1934.

A considerable number of people were now financially dependent on Stravinsky; and remembering his wartime difficulties, when he had been unable to rely on a regular income as a composer, he decided he must diversify his working life. This meant devoting time to carrying out engagements as a pianist and conductor at the expense of composing. The first noticeable effect of this decision was the appearance of a number of new works for piano. At Anglet in 1921 he had made a virtuoso piano transcription of three movements from *Petrushka* for Artur Rubinstein. Now he saw to it that his new piano works fitted his own limited executive capacity. He wrote the Concerto for piano and wind instruments (1923–4) and the Capriccio for piano and orchestra (1928–9) with this requirement in mind, retaining in each case the exclusive rights of performance for a period of five years. Later, when his younger son Sulima became a professional pianist, he wrote a Concerto for two solo pianos (1931–5) which was widely performed by father and son. He also wrote two solo pieces, the Sonata (1924), which he played at the ISCM Festival in Venice in summer 1925, and the Serenade in A. His 1924 concert tours took him to over a dozen European towns; and in 1925 he undertook his first American tour, scoring an undoubted success in his dual role of conductor and pianist. This visit led to his being commissioned by a gramophone company to write a piano work in four movements, each of which would fill one side of a ten-inch record, and the result was the Serenade, which he dedicated to his wife.

Stravinsky now had less time to devote to the Ballets Russes. It was true that Dyagilev revived Stravinsky's pre-war ballets and in 1923 gave the long delayed first performance of *The Wedding*. He also mounted productions of *Reynard* and *Mavra* (though not *The Soldier's Tale*). But after *Pulcinella* Stravinsky accepted no more commissions from the Ballets Russes. Instead he planned a special tribute to commemorate the 20th anniversary (in 1927) of the launching of Dyagilev's great theatrical enterprise in Paris and western Europe. This was the composition of *Oedipus rex*, which he cast in the form of an opera–oratorio, using as text the Latin translation of a libretto by Jean Cocteau. Because of production difficulties both Stravinsky and Dyagilev had to be content with a concert performance of *Oedipus rex* mounted in the same programme as a stage performance of *The Firebird*. Dyagilev was reserved in his attitude to *Oedipus rex* – 'un cadeau très macabre' he called it – and the majority of the audience did not like it. It needed a good stage production before its great qualities could be appreciated. In 1928 Stravinsky allowed Dyagilev to acquire the European rights of *Apollon musagète* (later renamed *Apollo*), which had been commissioned by Elizabeth Sprague Coolidge for production at the Library of Congress, Washington, DC. But shortly afterwards, to show that he no longer felt exclusively bound to Dyagilev and the Ballets Russes, even in Europe, he accepted a commission from Ida Rubinstein, who was forming a new ballet company of her own; and *Le baiser de la fée* ('The Fairy's Kiss', 1928), based by Stravinsky on songs and piano pieces by Tchaikovsky, was produced at the Paris Opéra (27 November 1928). Dyagilev was scathing about what he considered to be Stravinsky's defection from his fold; but there was not sufficient time for the breach to be healed, for on 19 August 1929 Dyagilev died in Venice and was buried on the island of S Michele. His death was followed by the immediate disbandment of his company.

In the mid-1920s Stravinsky experienced a spiritual crisis. He had been born and baptized into the Russian Orthodox Church, but at the age of 18 had abandoned the practice of his faith. In 1926 he rejoined the Orthodox Church, becoming a communicant once more; and the same year he had a profoundly moving religious experience when attending the 700th anniversary celebrations of St Anthony in Padua. This spiritual change affected some of his compositions. Between 1926 and 1934 he wrote three Slavonic sacred choruses, *Otche nash'* ('Our Father'), *Simvol' veri* ('Symbol of Faith', i.e. *Credo*) and *Bogoroditse devo* ('Blessed Virgin', i.e. *Ave Maria*); and when Koussevitzky asked him to write a major symphonic work to celebrate the Boston SO's 50th anniversary in 1930, he decided to compose a Symphony of Psalms to Latin texts selected from the Vulgate.

18. Stravinsky at the Café Gaillac, Paris, 1923

In 1931 he was introduced to the violinist Samuel Dushkin by Willy Strecker of Schott, Mainz. Out of this meeting came the idea of a close cooperation between the two musicians. First, Stravinsky wrote the Violin Concerto in D (1931), which Dushkin played in many countries with Stravinsky conducting. They then decided to become recitalists, and Stravinsky wrote the *Duo concertant* (1931–2) for violin and piano. This became the major work in their recital programmes, which also included a number of special transcriptions of some of Stravinsky's shorter pieces.

In the mid-1930s it seemed that the French phase in Stravinsky's life was approaching a climax. He had never shown himself much addicted to the language – so far his only setting of a French text had been the Verlaine songs of 1910 – but now a further ballet commission from Ida Rubinstein led to a collaboration with Gide. The proposal was that Stravinsky should make a musical setting of an early poem of Gide's, based on the Homeric hymn to Demeter. Stravinsky accepted; and his *Perséphone*, cast in the form of a musical melodrama, was produced at the Paris Opéra (30 April 1934). Gide seems to have been considerably upset by Stravinsky's 'syllabic' treatment of his text, which did not fit in with his strongly held views on the nature of French prosody; and the work had a rather muted reception, despite the express approval of at least one important critic, Valéry. On 10 June 1934 Stravinsky's naturalization papers came through, and he became a French citizen. The following year, reacting to the promptings of various friends (including Valéry), he sought election to the seat in the Institut de France left vacant by Dukas' death; but he was beaten in the poll by Florent Schmitt. This was the moment he chose to publish an autobiographical work in French, *Chroniques de ma vie* (1935). His first authorized publication had been an article on his Octet published in the periodical *The Arts* (January 1924); and thereafter he had issued a number of manifestos heralding each of his major new compositions. But *Chroniques de ma vie* was his first book as such; and in writing it he enjoyed the assistance of Walter Nouvel.

While failing to consolidate his position as a French artist in France, Stravinsky found an increasing demand for his music in the USA. In 1935 he carried out a second American tour, conducting many of the American orchestras and appearing with Dushkin at a number of violin and piano recitals. The following year he and his son Sulima played the Concerto for two solo pianos on a tour through South America. He then received an invitation from Lincoln Kirstein and Edward Warburg to write a ballet score to be choreographed by Balanchin for the recently formed American Ballet. The outcome was *Jeu de cartes* ('The Card Party'), composed in 1936 and first performed at the Metropolitan Opera House, New York (27 April 1937), with the composer conducting. Other American commissions followed. For Mr and Mrs Robert Woods Bliss he wrote a concerto for chamber orchestra (1937–8) which became familiarly known as the 'Dumbarton Oaks' Concerto after the name of their property in Washington, DC. Mrs Bliss was also largely responsible for obtaining for him a commission for a symphony to celebrate the 50th concert season of the Chicago SO

151

in 1940–41. About the same time an invitation reached him from Harvard University, offering him the chair of poetry for the academic year 1939–40.

These commissions and invitations came at a crucial moment. By now it was clear that Europe was poised on the brink of another world war; and at the same time ill-health (mainly tuberculosis) struck down several members of Stravinsky's family. His elder daughter, Ludmila, died in 1938 aged 30; his wife died in March 1939 aged 57; his mother three months later aged 85. He himself spent some time in the sanatorium where his wife and daughter had been; but by September 1939, shortly after the outbreak of war, he was sufficiently recovered to embark for New York on his fourth and what was to prove his longest visit to North America.

II Works

Although *Pulcinella* had pointed the way forward to a more universal musical style, Stravinsky's first major compositions after the end of the war constituted rather a final sublimation of his national idiom. Both the Concertino for string quartet (so called because of the prominence of the first violin part) and the Symphonies of Wind Instruments contain climactic sections in the 'Russian dance' style that goes back through the first of the Three Pieces for string quartet to *Petrushka*; the Symphonies, moreover, make use of the two other cardinal elements isolated in the string quartet pieces, developing them and bringing them into a new and completely original synthesis by means of metrically geared juxtaposition; the final chorale which crowns the work (the original memorial to Debussy) is the first fully developed example of the Stravinskian codas already referred to. Undervalued for many years, the Symphonies of Wind Instruments have eventually been recognized as a landmark in Stravinsky's output and have exercised a seminal influence on composers since World War II – notably on Stockhausen, of whose 'moment form' it has been proclaimed (however implausibly) a precursor.

The primacy of the 18th century (interpreted as including the last years of its predecessor) or Baroque music (interpreted as including Haydn and Mozart) as a stylistic persona for Stravinsky during his 'neo-classical period' should not blind one to the fact that he occasionally felt drawn to other idioms, among them those which formed part of his personal Russian heritage – not the atavistic heritage of folk melos which he had so fruitfully explored already, but that of Russian 19th-century art music. The little comic opera *Mavra*, scarcely more than an extended single scene with contrasting episodes, is dedicated to Pushkin (from whom the story is taken), Glinka and Tchaikovsky. The musical idiom makes more reference to Glinka, with his characteristic mixture of Italian and Russian turns of phrase, than to Tchaikovsky, who was to receive his own individual tribute in *The Fairy's Kiss* six and a half years later. Both the harmonic and the rhythmic aspects of *Mavra* show a new lucidity, and this is reflected in the instrumentation for a sizable wind ensemble (cf the Symphonies) with only a small complement of strings. This reversal of the traditional orchestral balance is carried through into the internal balance of the string section (the score calls for two violins, one viola, three cellos and three basses), and it reflects Stravinsky's total rejection at this period of the expressive excesses, as he and many of his generation felt them to be, of the symphonic composers of the German Romantic tradition. That this rejection has a great deal to do with the traumatic experience of World War I is obvious, though more purely musical reasons can also be found for it. At any rate the clarity of timbre and attack characteristic of the wind band predominates in other Stravinsky scores of the early 1920s, notably the Octet and the Piano Concerto. It represents a slight enrichment of the black and white palette of *The Wedding*, but only with the opera–oratorio *Oedipus rex* (1926–7) did Stravinsky work his way back to full acceptance of the symphony orchestra and the traditional role of the strings in it – an acceptance which was then celebrated in *Apollo*, for strings only. In this ballet, a hymn to the concepts of reason and enlightenment as the foundation of beauty, Stravinsky rejoices in the sonority of strings, augmented in this instance by a divided cello section, as completely as he had once rejected it. But it must be emphasized that he remained throughout his career an empiricist in the matter of instrumentation. While accepting the general outline of the symphony orchestra until the end of his neo-classical period, he varied it in numbers and in detailed composition. Each work was seen as a fresh problem in finding the appropriate, elegant and economical sound either for the external circumstances (as in the *Ebony Concerto*) or more often for the inherent needs of the musical idea.

There is an element of affectionate parody in Stravinsky's use of Russian materials in *Mavra* and *The Fairy's Kiss*, but his use of 'classical' styles evidently reflected a still deeper need, since it contained no admixture of irony, at least at first. For Stravinsky the period of music, and more generally of culture, bounded by Lully and Mozart represented an ideal of civilization, which attracted him both as a Russian, and thus as an alien to the charmed circle of nations for whom it was a part of their heritage, and also as a refugee from an at least theoretically egalitarian political system. The style of his middle years is thus quite literally and consciously reactionary, and has been rejected as such by those whose allegiance is to a Romantic and often doctrinaire notion of artistic progress. What Stravinsky's style is not, however, is conservative, and indeed his willingness, at this period, to retain the manner of a traditionalist while totally rejecting the matter, has proved a stumbling-block to genuine conservatives in their turn. The justification of this 'neo-classical' idiom lies, in any case, in the works which embody it. Stravinsky's adoption of some mannerisms of 18th-century music, together with his maintenance of the notion of related centres of tonal gravity even while rejecting the traditional means of establishing them, provided him with a language capable of many very different kinds of utterance, which he exploited fully in the interwar years. The lyrical beauty of the Octet, the Piano Sonata and Serenade, and above all the *Duo concertant,* apart from works already mentioned, would have been impossible without what Eric Walter White has called, in a memorable phrase, Stravinsky's 'sacrifice to Apollo'. At the same time his acceptance of neo-Baroque formal disciplines also made it possible for him to compose masterpieces with the specific gravity of *Oedipus rex*, the Symphony of Psalms, and the balletic retelling of the regeneration-mystery, *Persephone*.

All three of these great works are, in their distinct

ways, eclectic in style but highly original in form and tone. In *Oedipus rex* and *Persephone* Stravinsky makes prominent use of the spoken voice in conjunction with music, though to quite different effect. In the former the use of a narrator, speaking the audience's language, to introduce the scenes of a drama composed in the liturgical 'dead' language, Latin, strengthens their sense of monumentality. In *Persephone*, however, the performer of the title role, intended for Ida Rubinstein, declaims Gide's flowery text over the music, where the other characters sing it; this introduces a weakening shift of dramatic focus, as well as making the composed (most unusually in Stravinsky) serve as a background to the uncomposed. Both works – the one a tragedy, the other a mystery – convey an almost ritual weight of utterance, but this is most consistently felt in the Symphony of Psalms. The text is drawn from verses of the Vulgate version, arranged (as in Stravinsky's later religious works) to form a highly characteristic sequence of repentance, faith and praise. Characteristic too is the deliberate avoidance of self expression (the reverse of Beethoven's or Bruckner's religious music) and the cultivation of an impersonal, objective liturgical persona. The work is a symphony only in the loosest of senses, and may have been so called only because the Boston SO had commissioned a symphonic work. Not for another decade would Stravinsky come to terms with the traditional symphony, as represented by Haydn and Beethoven, in the Symphony in C. Insofar as the Symphony of Psalms refers to the past at all, it is to the choral works of the Baroque period, in which chorus and orchestra are used on an equal footing for contrapuntal development. The composition of the orchestra also reflects Stravinsky's desire for an archaic monumentality: not only are violins and violas absent, but also those 'modern' instruments, the clarinets. The wind section is large, however, and two pianos are prominent among the percussion.

After *Persephone* it must be admitted that some flagging of creative impulse can be felt in Stravinsky's works of the later 1930s. The Concerto for two solo pianos, in which he again tackled the challenge of abstract composition without the stimulus either of stage action or of instrumental colour, is a tough, impressive, but not entirely convincing work; *The Card Party* (1936) is a brilliant and inventive ballet score, but without the powerful resonance of *Apollo* and its predecessors, and the 'Dumbarton Oaks' Concerto, though witty and resourceful in its evocation of Bach's Brandenburg manner, remains a minor work. Family tragedies, political anxieties, his own ill-health and, perhaps more important, a feeling that his music was insufficiently recognized in his adopted country, probably all took their toll of Stravinsky's vitality at this time. Thus his second emigration, to the USA, though it was undoubtedly disturbing and unsettling, may well have helped to bring about the eventual renewal of his creative activity.

CHAPTER FIVE

1939–52

I Life

Stravinsky landed in New York on 30 September 1939
and went straight to Cambridge, Massachusetts. His
six Charles Eliot Norton lectures on the poetics of
music had been written in French with the aid of Roland-
Manuel and Pierre Souvtchinsky before he left Paris, and
he delivered them (also in French) to large audiences in
Harvard's New Lecture Hall. In 1942 the original text
was published by the Harvard University Press as
Poétique musicale.

In December 1939 Stravinsky conducted concerts in
San Francisco and Los Angeles, returning to New York
in time to meet Vera de Bosset when she arrived (on 13
January 1940) by sea from Genoa. On 9 March they
were married in Bedford, Massachusetts, and in May
they went to Galveston and Houston on an ill-planned
honeymoon trip. Because he had found the Californian
climate suited him, they decided to settle in Hollywood.
In July that summer they applied for visas and went to
Mexico to establish quota qualifications and, re-entering
the USA as Russian non-preference quota immigrants,
they filed declarations of intent to become American
citizens.

When Stravinsky had reached America only the first
two movements of the symphony intended for the

Chicago SO were complete. The third was written during the period of his Harvard lectures, the fourth in Hollywood in summer 1940; the first performance of the Symphony in C was given that autumn in Chicago with the composer conducting.

Once he had settled in Hollywood Stravinsky was inevitably approached with suggestions that he should write music for films. Some of these projects were so fatuous they were dismissed at once; but others engaged his interest, and in a few cases he had begun to compose music before they were abandoned. Fortunately he was able to salvage a considerable part of this abortive film music and use it again in such works as *Four Norwegian Moods* (1942), the *Ode* (1943), the *Scherzo à la russe* (1944) and the Symphony in Three Movements (1942–5).

Shortage of cash may have played some part in leading Stravinsky for the first and only time in his life to accept a private composition pupil: Ernest Anderson went to him in March 1941 and received approximately 215 lessons during the next two years. The same factor may also help to explain the alacrity with which he accepted some slightly offbeat commissions – particularly the *Circus Polka* (1942) to be danced by young elephants in the Barnum and Bailey Circus, and *Babel* (1944) which formed part of a composite biblical cycle called *Genesis*, commissioned by Nathaniel Shilkret, in which seven different composers (including Shilkret himself) were involved. Other wartime commissions included *Danses concertantes* (1941–2), a concert score written for the Werner Janssen Orchestra of Los Angeles, though later it was used for ballet purposes, and *Scènes de ballet* (1944) written for a ballet that was incorporated in Billy Rose's Broadway revue *The Seven Lively Arts*.

The most important work of this period was the result of a commission from the New York Philharmonic Symphony Society, which arrived when Stravinsky was contemplating the possibility of writing another piano concerto. In the event, much of the music sketched out for this concerto was incorporated in the first movement of the Symphony in Three Movements, while some of his abortive film music was used in the second movement. The Symphony in Three Movements made a powerful impression when it was first performed by the New York PO on 24 January 1946.

Stravinsky and his wife were now American citizens, their naturalization papers having gone through on 28 December 1945. This gave him a chance to review his work to date and see what he could do to safeguard his earlier copyrights. His early music was not protected in the USA, as at the time it was written Russia and the USA had not ratified the Berne copyright convention. Later he had tried to safeguard the copyrights of his new compositions by the subterfuge of appointing an American editor; but when he became a French citizen in 1934 he found that all his works from 1931 onwards became automatically protected in the USA. At the end of 1945 he signed an agreement with Ralph Hawkes of Boosey & Hawkes giving that firm exclusive publication rights for all his future compositions; and at the same time they took over those earlier works of his, ranging from *Petrushka* to *Persephone*, that had formerly been published by the Editions Russes de Musique. (This did not affect works of his published by other firms, such as

Chester and Schott.) Now it became possible for him to make new versions of his earlier Editions Russes scores, to revise his music where necessary, to correct the numerous errors that marred some of the early editions, and to make sure that the copyright of these revised versions was properly protected. Some of his revisions were extensive. He rescored *Petrushka* for smaller orchestra (1946); and the make-up of the orchestra was changed in the new version of the Symphonies of Wind Instruments (1947). Less radical changes were made in the revised versions of *Apollo* (1947), the *Pulcinella* suite (1947), *Oedipus rex* (1948), the Symphony of Psalms (1948), the concert suite from *The Fairy's Kiss* entitled 'Divertimento' (1949), the Capriccio (1949), *Persephone* (1949), the Concerto for piano and wind instruments (1950), *The Fairy's Kiss* (1950), the Octet (1952) and *The Nightingale* (1962).

The end of the war brought Stravinsky a commission from Europe, the first for over ten years. This was for the Concerto in D for string orchestra, which he wrote in 1946 at the behest of Paul Sacher for the Basle Chamber Orchestra. It was followed by *Orpheus*, a new ballet score, commissioned by Lincoln Kirstein for the Ballet Society, which produced it in 1948 with choreography by Balanchin in New York.

Since *Oedipus rex* it had become exceedingly rare for Stravinsky to write a non-commissioned work. In 1942, however, he decided for personal reasons that he wanted to write a mass, a 'real' mass, and by that he meant a Roman Catholic mass that could be used liturgically. The initial impetus came from his discovery of some Mozart masses in a second-hand music store in Los Angeles; and he planned his mass for mixed chorus and double wind quintet. The Kyrie and Gloria were written in 1944; the Credo, Sanctus and Agnus Dei followed in 1947 and 1948. The first performance was given at La Scala, Milan, on 27 October 1948 with Ansermet conducting. Performances of the Mass as part of a church service have been comparatively rare.

Stravinsky now made an important decision – that he would compose a full-length opera in English. Although the work had not been commissioned by any specific person or institution, his relations with Boosey & Hawkes were so satisfactory that he felt justified in deciding to devote three years of his composing life to this opera. This would also be his first full-length work for the stage, for hitherto none of his operas or ballets had played for more than an hour. In the course of a visit to the Chicago Art Institute in 1947 he had been greatly impressed by a set of Hogarth's engravings depicting *The Rake's Progress*. Here Stravinsky thought was material for an opera libretto, and on the advice of his friend and neighbour Aldous Huxley he approached Auden and invited him to visit him in Hollywood. The meeting was successful; and, with the collaboration of Chester Kallman, Auden was able to hand over the completed libretto in Washington, DC, on 31 March 1948. Each of the three acts took roughly a year to write, and the completed opera was ready by April 1951. The first performance was given in Venice on 11 September 1951 at La Fenice during the 14th International Festival of Contemporary Music. La Scala supplied the chorus and orchestra, while the principal roles were cast from the best singers available; the composer himself conducted.

The Rake's Progress was an undoubted success and

firmly re-established Stravinsky's reputation in postwar Europe. What was not realized at the time, however, was that in perspective it would be seen not only as the climax of his neo-classical period, but also as its peroration. Shortly afterwards Auden offered him another libretto to set, this time a masque entitled *Delia*; but by temperament Stravinsky was never inclined to run the risk of repeating himself. Instead, on returning to the USA, he chose four early English anonymous lyrics from an anthology edited by Auden and Norman Holmes Pearson and set them for soprano, tenor, chorus and small instrumental ensemble as the Cantata (1951–2). This work received its first performance by the Los Angeles Symphony Society in autumn 1952.

II Works

Stravinsky's removal to the USA did not bring about any immediate change in his style, but this is not altogether surprising when one considers that his professional activities there had been increasing since the mid-1930s. Although the first two movements of the Symphony in C were composed in Europe it is hard to agree with Stravinsky's own assertion that there is a marked stylistic difference between them and the last two. It was at this time that he gave, in the Harvard lectures, the definitive account of his aesthetic stance; whether the stimulus was external, from the change of scene, or internal, from a new consciousness of the challenge of symphonic form, the Symphony in C represents an intensification of his neo-classicism. The work places great emphasis on motivic relationships, but it also makes deliberate and effective use of delayed or contradicted expectation – not only on the foreground level of rhythmic and harmonic detail, but on the larger formal level as well. This is not the wit of a Haydn or a Beethoven, playing with conventions they essentially accept; Stravinsky's attitude towards tonality, his means of establishing and relating tonal centres, are quite different from those of the Baroque and Classical composers whose manners and mannerisms he occasionally saw fit to borrow. The controlled dislocation he cultivated here, beneath an urbane exterior, reveals a sensibility that rejects the banal without placing any reliance on spontaneous expression.

That a subtly ambiguous form of expression can nevertheless be achieved by these means is proved above all by the opera *The Rake's Progress*, the summa of Stravinsky's neo-classical period, in which he was fortunate enough to be abetted by a librettist whose sensibility was completely in tune with this aspect of his own; there, at last, the irony latent in the style is for the first time exploited. But although these two works, the Symphony in C and the opera, mark the beginning and end of the first period of Stravinsky's life in the USA, they are far from exhausting its musical character. From the first years after his settlement in Hollywood, when money was a problem and commissions a permanent temptation, there date a number of frankly lightweight works, such as the *Circus Polka* and the *Scherzo à la russe*; others contain uncomfortable variations of tone which perhaps reflect his unsettled state and his emotional involvement in the war. If the final apotheosis of *Scènes de ballet* seems suddenly to be charged with an intensity quite out of scale with what has gone before, or with its function in a Broadway revue, this is no doubt due to the fact that Stravinsky received the news of the liberation of Paris while he was at work on it, as he noted in the score. Similar considerations may account for the startlingly Dionysian quality of parts of the Symphony in Three Movements, composed only five years after the Symphony in C, but utterly different from it in mood and even in technique. Where the earlier work seems to have been conceived from the first as a lucid and clearsighted essay in the mode of the classical symphony, the later one is rather a concatenation of various ideas. Some of them were suggested (according to Stravinsky) by images in wartime newsreels; the middle movement, however, was originally composed to accompany the apparition of the Virgin in the film of Franz Werfel's *Song of Bernadette*, and its connection with the other two is purely one of contrast. The work's coherence comes less from Stravinsky's highly accomplished compositional craft than from the inherent vitality of its constituent ideas; it is this that transcends the work's abrupt transitions and the inconsistency of its textures, and makes of it a more popular, though perhaps less perfect, score than the Symphony in C.

As if in reaction to this Dionysian upsurge, a new austerity made itself heard in two immediately postwar works, though both also suggest that Stravinsky was beginning to cast his net wider in terms of historical reference. In the Mass liturgical function imposes a certain simplicity of style, but within that the voices (which are kept in the forefront throughout) borrow gestures from 14th-century Ars Nova and 16th-century polyphony as well as from the syllabic chant characteristic of Stravinsky's Orthodox church music. In the ballet *Orpheus* the concertante use of the harp may have been suggested by a similar feature in the central aria of

Monteverdi's operatic treatment of the same myth. Otherwise its stylistic links are mostly within Stravinsky's own music: the second act of *Persephone* (with which it is connected by its subject matter) and, in the dance in which the Bacchantes tear Orpheus to pieces, the violent sections of the Symphony in Three Movements. Certain sections, notably the two thematically related fugal interludes and the final apotheosis, with its free canonic writing for two horns, betray a new interest in counterpoint that looks forward to Stravinsky's last creative period.

Before that, though, there came the amazing achievement, utterly unpredictable in its youthful exuberance, of *The Rake's Progress* – at about two and a half hours Stravinsky's longest work, though it was completed in his 69th year. The fluency with which the work was composed owed much to the mutual respect and sympathy between Stravinsky and Auden (though he was assisted by Chester Kallman, Auden's was clearly the dominant poetic character). The subject itself suggested an 18th-century frame of musical reference, and this found an immediate response in Auden's own, more amateur, enthusiasm for the older opera (unlike Stravinsky he was also a keen Wagnerian). The artistic development of both men had led them to a willing, if individual, acceptance of convention as a defence against chaos and formlessness; they shared, too, a need to screen the unconscious springs of their inspiration behind a scrupulous preoccupation with craftsmanship. Thus *The Rake's Progress*, in spite of its use of arias and recitatives (dry and accompanied), of ensembles and choruses, not to mention the 18th-century turns of expression in both music and text, is far from being a mere pastiche. Indeed the very strictness of the conventional framework seems to have liberated an unusual directness of emotional expression in Stravinsky. Like all his major theatrical works, *The Rake's Progress* is as much fable as story; it demands both empathy with the characters and an awareness of their meaning. The story, though loosely based on Hogarth, combines at least two of Stravinsky's main dramatic preoccupations: the conflict of good and evil in the human soul (as in *The Soldier's Tale*) and the cyclical regeneration of nature seen as a metaphor for that of mankind (the seasonal framework which he had suggested for *Persephone* is here made explicit). The vaudeville epilogue, for all its apparent flippancy, is much more than an antiquarian nod to *Don Giovanni*. 'For idle hands/ And hearts and minds/The Devil finds/A work to do . . .'; it is a succinct and witty statement of the philosophical viewpoint that underlies both Stravinsky's and Auden's sustained creative activity.

CHAPTER SIX

1953–71

I Life

In the summer of 1947 Robert Craft, then a young man of 23, wrote to Stravinsky from New York asking if he might borrow a copy of the score of the Symphonies of Wind Instruments (then unobtainable) for a concert of the Chamber Art Society, New York, which he was to conduct. The correspondence led to a generous offer by Stravinsky – that he was ready himself to take part in the concert by conducting the Symphonies and *Danses concertantes* without fee, leaving Craft to conduct the Capriccio and the Symphony in C. At the end of March 1948 Craft travelled to Washington, DC, to arrange details and was introduced to Stravinsky and his wife by Auden, who happened to be delivering the final text of the libretto of *The Rake's Progress* at the same time. The meeting was a success, and later Craft was invited to stay with the Stravinskys in Hollywood, where he helped the composer by cataloguing a substantial batch of his music manuscripts that had just arrived from Paris and answering various queries that arose from his setting of Auden's libretto. The friendship prospered; and Craft's position as musical aide and assistant was confirmed when it was realized that there were many ways in which he could be of continuing help to the composer, now in his 70s.

Craft, who had always been interested in the music of

the Viennese serialists as well as that of Stravinsky, was at first surprised to find that in Hollywood Stravinsky and Schoenberg, who lived within a few miles of each other, kept strictly to themselves. After Schoenberg's death (13 July 1951) he encouraged Stravinsky to listen to recordings of a wide range of serial music, and the composer found himself growing receptive to the music of Webern. His own approach to serialism as a composer was a cautious one. In May 1952, when visiting Paris, he still maintained in replying to a journalist's question that the serialists were prisoners of the number 12, while he felt greater freedom with seven. Nevertheless, this was the moment when he began to experiment with some of the processes used by the serialists.

The world-wide success of *The Rake's Progress* meant that sooner or later Stravinsky was bound to be asked to write another opera; and in 1953 Boston University offered him such a commission. The Stravinskys had been favourably impressed when they heard Dylan Thomas give one of his poetry readings at Urbana in 1950; now he was invited to cooperate with Stravinsky as librettist. The two men met in Boston in May 1953 and liked each other on sight; it was agreed that Thomas should visit the Stravinskys in Hollywood that autumn in order to draft a suitable scenario. Unfortunately this visit never took place, for Thomas died in New York on 9 November. The opera project lapsed; instead Stravinsky composed an elegy, *In memoriam Dylan Thomas* (1954), in which a setting of Thomas's poem 'Do not go gentle into that good night' for tenor and string quartet is framed by dirge canons for a quartet of trombones.

The organizers of the Venice Biennale now commissioned a new work from Stravinsky for 1956. Wishing to make this work particularly Venice's own, he decided to inscribe it to Venice's patron, St Mark, and design it for performance in St Mark's Cathedral. The *Canticum sacrum*, for tenor and baritone, chorus and orchestra, was a setting of various passages from the Vulgate arranged in a cycle of five movements. As the work lasted for only 17 minutes – one of the characteristics of Stravinsky's new serial idiom being its increasing compression – he originally intended the programme should include a recomposed work by Gesualdo as well; but the Venetians refused to admit the music of a Neapolitan in St Mark's, so he offered an instrumental arrangement of Bach's canonic variations on *Vom Himmel hoch, da komm' ich her* instead. The première of the *Canticum sacrum* was given in Venice on 13 September 1956 with the composer conducting.

1957 saw the completion of a new ballet score, *Agon*, which had been commissioned by Lincoln Kirstein and Balanchin as the result of a grant made to the New York City Ballet by the Rockefeller Foundation. Stravinsky found a prototype for some of his dance movements in Lauze's *Apologie de la danse* (1623) and Mersenne's music examples. The composition of this score started as early as December 1953, but was interrupted by the need to complete *In memoriam Dylan Thomas* and the *Canticum sacrum*. When Stravinsky returned to *Agon* in 1956 he found he had to recast some of the early numbers to link them more sympathetically to his serial music, the technique of which was now fully developed. The first concert performance of the new score was given on 17 June 1957, when it was conducted by Craft as part of a special Los Angeles

170

festival programme to commemorate Stravinsky's 75th birthday. The first stage production was given by the New York City Ballet on 1 December 1957.

In connection with Stravinsky's 75th birthday Craft had the idea that it might be helpful publicity (and save the composer the fatigue of being interviewed by too many reporters) if he himself interviewed Stravinsky and published the resultant text with the composer's approval. He did so; and 'Answers to 35 Questions' duly appeared in numerous periodicals. As this particular formula appeared to work satisfactorily, the two authors became more ambitious. Other questions were asked; other answers given; opportunities were found to include some of the correspondence Stravinsky had received from friends at various stages of his life; and *Conversations with Igor Stravinsky* was published in 1959.

Other volumes compiled by the two collaborators followed, *Memories and Commentaries* (1960), *Expositions and Developments* (1962), *Dialogues and a Diary* (1963), *Themes and Episodes* (1966) and *Retrospectives and Conclusions* (1969). As this collaboration progressed the formula for compiling the books started to change. In the first place, the last three volumes contained substantial extracts from Craft's diaries, which were later reprinted in his book *Stravinsky: Chronicle of a Friendship 1948–1971* (1972). Then it appeared that the two authors were beginning to sink their individual identities in a new character which was distinguished by some of the salient characteristics of both.

A further commission for Venice was received in 1957. This time it came from North German Radio,

19. *Stravinsky in London, 1958*

whose orchestra and chorus gave the first performance of the cantata *Threni* in the Sala della Scuola Grande di S Rocco on 23 September 1958 as part of the Venice Biennale programme. *Threni* was followed by Movements for piano and orchestra (1958–9), written for Margrit Weber; and a further commission from Paul

Sacher resulted in *A Sermon, a Narrative and a Prayer* (1960–61).

During the war Stravinsky had accepted a number of conducting engagements, mainly in the USA. His return to Europe in 1951 at the time of the first performance of *The Rake's Progress* had led to a renewal of invitations to conduct overseas, mainly in Italy, Germany and Switzerland, and occasionally in Paris and London. He now discovered that Craft could be of great help to him by preparing and rehearsing the orchestra before he took over, and sometimes also by sharing the burden of conducting. The fact that Craft was able and willing to accompany Stravinsky and his wife made these trips more agreeable and less onerous. From 1958 for a period of about ten years the number of these tours increased enormously; and, instead of being confined to North America and Europe, they now spread all over the world – to South America, the Far East, Australasia and Africa. In the 1950s Stravinsky also entered into a contract with Columbia under which all his works were to be recorded with the composer as conductor. Here too Craft was invaluable in helping rehearse the orchestra before Stravinsky took over for the final run-through and recording.

Plans to celebrate Stravinsky's 80th birthday in 1962 proceeded on a much more extensive scale than those for his 75th. On 16 January he received the State Department's medal, and two days later he and his wife were guests of the President and Mrs Kennedy at a dinner party in the White House. The following month a new anthem to words by Eliot, 'The dove descending breaks the air', was given at one of the Los Angeles

Monday Evening Concerts; the same month *A Sermon, a Narrative and a Prayer* had its première in Basle; and the television broadcast of *The Flood* by CBS, who commissioned the work, took place on 14 June. Stravinsky celebrated his 80th birthday in Hamburg, where a special programme consisting of *Orpheus*, *Agon* and *Apollo*, all with choreography by Balanchin, was mounted by the New York City Ballet.

Undoubtedly the most important event of 1962, however, was Stravinsky's return to Russia after an absence of nearly half a century. The previous year a deputation of Soviet musicians had transmitted to him in Hollywood an official invitation to visit the USSR and conduct a concert of his own music on the occasion of his 80th birthday. A number of his friends and acquaintances thought that for political and other reasons he should refuse; but in the end saner counsels prevailed, and he agreed to give a series of concerts in Moscow and Leningrad. The Russian visit came at the end of an extensive tour embracing Toronto, Paris, Brazzaville, Johannesburg, Pretoria, Cape Town, Rome, Hamburg and Israel. The Stravinskys, together with Craft, arrived in Moscow on 21 September. Three concerts were given there, and two in Leningrad; and the intervening days were passed in a whirl of sight-seeing and lavish entertainment. On their last day in Moscow (11 October) they were received by Khrushchev in the Kremlin. The visit was undoubtedly a great success. It marked the beginning of a more liberal attitude to Stravinsky's music in the USSR; and, for the composer himself, Craft perceptively noted in his diary, 'To be recognised and acclaimed as a Russian in Russia, and to be performed there, has meant more to him than anything else in the years I have known him'.

Stravinsky's next composition was a work commissioned by the Israel Festival Committee, the sacred ballad for baritone and small orchestra *Abraham and Isaac* (1962–3). He set the text in the original Hebrew; which brought the number of languages he had set to seven, the others being Russian, French, Italian, Church Slavonic, Latin and English. The first performance of *Abraham and Isaac* took place in Jerusalem on 23 August 1964.

As Stravinsky's life lengthened many of his friends and acquaintances died, and occasionally he felt prompted to write some kind of musical epitaph. *In memoriam Dylan Thomas* belongs to this category; so too do two miniature serial works, both composed in 1959, both lasting only just over a minute, and both creating an effect of classical summation – the *Epitaphium* for flute, clarinet and harp 'für das Grabmal des Prinzen Max Egon zu Fürstenberg' and the Double Canon for string quartet 'Raoul Dufy in memoriam'. The assassination of President Kennedy on 22 November 1963 led Stravinsky to write a miniature *Elegy for J.F.K.* for baritone and three clarinets (1964) to four stanzas specially written by Auden, each a 'free' haiku. On the day of Kennedy's assassination Aldous Huxley also died, and Stravinsky decided to dedicate to his memory a work he had already started to compose, the Variations for orchestra.

It is probable that towards the end of 1964 Stravinsky had in mind the idea that he might compose a Requiem mass; but the death of his friend Eliot in

London on 4 January 1965 precipitated the composition of a single movement, *Introitus*, which was completed six weeks later. A commission from Princeton enabled him to give more attention to the requiem idea. It was a condition of the commission that the work should be dedicated to the memory of Helen Buchanan Seeger; but in reality the *Requiem Canticles* (1965–6) were written with his own approaching death in mind. This proved to be Stravinsky's last major composition. Shortly after its first performance at Princeton University (8 October 1966) he completed a lightweight setting of Lear's *The Owl and the Pussy-cat* for soprano and piano, which he dedicated to his wife. Early in 1968 he started to compose an extra instrumental prelude to the *Requiem Canticles* for a special performance of the work in memory of Martin Luther King; but he was unable to complete this in time. He also drafted some sketches for a piano sonata; but these were ultimately abandoned. An instrumental transcription of two sacred songs from Wolf's *Spanisches Liederbuch* was made in San Francisco in May 1968 and performed in Los Angeles later that year; but instrumental transcriptions of two preludes and fugues from Bach's '48' were withdrawn by Craft from a concert in Berlin in October 1969.

By 1967 Stravinsky's health was beginning to fail. In January he made his last recording (of the 1945 suite from *The Firebird*); and in May he conducted his last public performance (of the *Pulcinella* suite, in Toronto). After that he was occasionally 'in attendance' at concerts of his works conducted by Craft. He was tended devotedly by his wife and various nurses; after his composing faculty had started to fail Craft was successful in getting him to spend more time listening to the recorded music of other composers – particularly Beethoven – which gave him much pleasure. In 1969 the Stravinskys decided to move from Hollywood to New York; the following year they flew to Europe in the summer and spent three months at Evian on Lake Geneva. On 6 April 1971 Stravinsky died at his home in New York. At his widow's suggestion he was buried in Venice on the island of S Michele, not far from Dyagilev's grave.

II Works

Various circumstances combined to produce, in the years immediately following the composition of *The Rake's Progress*, the most profound change in Stravinsky's musical vocabulary that it had undergone for more than 30 years. One was the adoption into his family circle of Craft, whose enthusiasm for the music of the Second Viennese School certainly helped to focus Stravinsky's attention on it. Another was his return to Europe for the première of *The Rake* in 1951, and the contacts that this brought with a new generation of European composers, many of them strongly influenced by the postwar rediscovery of Schoenbergian serialism. And perhaps most important of all there was the death, earlier in 1951, of Schoenberg himself, for many years Stravinsky's fellow exile and neighbour (virtually unacknowledged) in Hollywood. The post-Romantic characteristics of Schoenberg's music and the doctrinaire cast of his thought, not to mention the prickliness of his personality, had prevented Stravinsky from coming to terms with him while he lived, and the two men had been cast by their respective followers as opponents in a quasi-ideological battle of style comparable to that of Wagner

177

and Brahms two generations earlier. As Stravinsky now gradually began to come to terms (his own, it must be said) with some aspects of serial practice, and moved further away from the tonal waters in which his music had hitherto sailed under a flag of convenience, many of his own most embattled followers undoubtedly felt a sense of betrayal. Yet in fact Stravinsky's 'conversion', unlike that of some other composers at about this time, was very far from being a capitulation. It was much more like an annexation of the enemy's resources to his own perennial purposes, and it was carried out with immense caution, each stylistic step being tested by the only criteria he had ever admitted, his instinct and his ear.

There were already precedents in Stravinsky's own music for treating a melody as a series (i.e. as a sequence of pitches with no inherent rhythmic or harmonic implications); examples occur in the variations of the Octet (1923), the last two movements of the Concerto for two pianos (1935) and the interludes in *Orpheus* (1947), in all of which the resulting series is, significantly enough, sooner or later treated fugally. Stravinsky now began to explore the spatial, purely intervallic implications of this practice by applying to such series the basic serial transformational procedures of retrograde, inversion and retrograde inversion. Thus in the middle movement of the Cantata on early English texts, a setting for tenor and instrumental quintet of the rhymed life of Christ 'Tomorrow shall be my dancing day', the solo line, with the exception of its punctuating refrain, is strictly derived from an initial sequence of 12 notes (as are some of the counterpoints to it). This is not an orthodox Schoenbergian series, since it involves only the seven pitches contained within an augmented 4th, and the serial procedures themselves do not penetrate all voices of the movement or all sections of the work, but it does involve a new emphasis on purely melodic aspects of development and integration. In the Septet (1953) for the Schoenbergian combination of three wind, three strings and piano, composed in E♭, like the 'Dumbarton Oaks' Concerto (1938), these procedures are taken further. Although the first movement is not far in style from its neo-Bachian predecessor, the second is a passacaglia on a 16-note theme, which provides, by serial and canonic manipulation, most of the contrapuntal texture around it, and the final gigue takes up the eight different pitches which the passacaglia theme contains and treats them as a transposable scale from which to derive a succession of fugue subjects; these too appear in retrograde and inverted form. The contrapuntal action in the Septet is increasingly dense and the harmonic movement increasingly rapid and dissonant. The Three Shakespeare Songs for mezzo-soprano, flute, clarinet and viola, and *In memoriam Dylan Thomas* for tenor and string quartet, with the addition of four funereal trombones (cf Schütz's *Fili mi Absalon* and Beethoven's *Equali*) in the canonic prelude and postlude, are further explorations of this essentially linear serialism, whose anti-tonal consequences are mitigated by the use of smaller pitch collections than the full spectrum of 12 and by a free use of transposition.

Just as the short vocal and instrumental works of 1914–16 have tended to be obscured by the larger works (*Reynard, The Wedding*) for which they were in some sense studies, so these short works of 1952–4 have inevitably been overshadowed by the larger and

179

more confident works that followed them; yet in spite of their rather cramped melodic style they are highly individual works in their own right. But with the *Canticum sacrum* a new breadth is immediately audible. Apart from the Ars Nova-like epigraph with which it begins, the work is strictly symmetrical, its groundplan apparently based on that of St Mark's. The Byzantine architecture may also have suggested the style of the choral first and fifth movements (the latter a palindrome of the former); they seem like a throwback to the Russian Stravinsky of *The Wedding* and the Symphonies of Wind Instruments. But the three middle movements – a lightly accompanied lyrical tenor solo, a solemn celebration of the three cardinal virtues of faith, hope and charity by the work's full forces, and an intensely dramatic setting for baritone and chorus of a prayer for faith drawn from St Mark's gospel – all these make use, with whatever concessions to Stravinsky's desire for tonal anchorages, of full 12-note series. That such a variety of textures and techniques is juxtaposed with no sense of incongruity is sufficient indication of the extent to which he had already succeeded in making over serial procedures to his own long-established musical purposes, but in *Agon* he achieved a more remarkable feat still: moving from diatonicism, through polytonality and partial serialism to a complete 12-note serialism (in the climactic pas de deux and the concerted dances that follow it), he then succeeded in returning without incongruity to the diatonic music with which the work began.

Threni, however, a setting of carefully selected portions of the Lamentations of Jeremiah for six solo voices, chorus and a rather large orchestra (including flügelhorn and sarrusophone but no trumpets or bassoons), is Stravinsky's first completely serial score. In the lightly scored *Agon* Stravinsky had come closest to Webern, whose pointillist, intervallic version of serial technique was far more congenial to him than Schoenberg's more harmonically orientated style. *Threni*, in keeping with its sombre text, is altogether more weighty in sound, and reveals very clearly the incompatibility of Stravinsky's harmonic sense with that of orthodox serialism. In the fuller sections repeated notes and phrases are constantly allowed to set up fields of tonal attraction, and although the more contrapuntal sections for smaller forces (notably the extraordinary set of unaccompanied canons for solo male voices in the 'Querimonia') avoid this almost completely, the music is finally brought, by cunning transposition of the various forms of the series, to a quasi-tonal cadence that sounds almost like A minor. Stravinsky's sense of the dramatic connotations of his text clearly demands a resolution of this kind, however contrary it may be to the theoretical principles that are supposed to underlie, and indeed justify, serial practice.

If the series remains, in *Threni*, an essentially linear entity, *Movements*, a brief but densely composed concert piece for piano and orchestra, achieves a new freedom by refracting the series into its constituent smaller motifs and recombining them in various ways – a form of serial punning. Rhythmically, too, *Movements* is far more flexible than *Threni*, or indeed any previous music by Stravinsky, but the work is given a firm framework by reserving a separate group of instruments to each section and by using the last few bars of each of its five brief movements (except the last) to introduce the tempo of the following one. *A Sermon, a Narrative and a Prayer* is to some extent a New Testament counterpart to *Threni*, but its texts take up the central burden of the *Canticum sacrum*, that of the three cardinal virtues: St Stephen, praying for his murderers in the moment of his martyrdom, is taken as the archetype of Christian love proceeding from faith and hope. Technically the work brings into Stravinsky's sacred music, especially in the central narrative of the martyrdom, some of the freedoms newly won in *Movements*, such as the flexibility of rhythm and the fragmentation of the orchestra's sonorities; this is extended to the voices, as when speaking and singing voices are ingeniously mingled, and when alto and tenor are dovetailed to form a composite solo voice of preternatural range. Only in the final section (almost the last, and certainly one of the greatest, of the long series of epitaphs with which Stravinsky had, throughout his life, commemorated the death of his friends) does the novel use of three tam-tams of different sizes to suggest the tolling of bells indicate a link with a much earlier Stravinsky.

In *A Sermon* Stravinsky's shaping genius is at its surest, placing and balancing the diverse textures and tempos with an unerring hand. In *The Flood*, a treatment of the Chester miracle play with additional material from the York cycle and elsewhere, commissioned by and designed for television, this large-scale architectural sense seems for once to have deserted him. The two danced sections – the building of the ark and the flood itself – show all his newly expanded mastery, but the remainder seems to contain music of too much diversity and not enough density to balance the lengthy spoken narration, in spite of such vivid and typical touches as the use of two basses in homophony to represent the impersonal voice of God (cf the bridegroom's request for a blessing in *The Wedding*), the insinuating high tenor for Satan (cf *Reynard*), the brass chords that accompany God's curse, and the woodwind and harp that depict the rainbow. Stravinsky's apparent attraction to themes already treated by Britten (a not altogether friendly emulation, one suspects) is carried a stage further by his setting of the story of *Abraham and Isaac*, where the sounds of the Hebrew text, as interpreted to Stravinsky by his friend Sir Isaiah Berlin, were the starting-point for the composition. The narrative given to the baritone soloist is kept in the forefront of the texture throughout, and the chamber orchestra (which surprisingly contains two trumpets, two trombones and tuba) is used with extreme economy; only at the climactic prophesy, in fact, that 'in thy seed shall all nations of the earth be blessed' are the above-mentioned brass used together, and then only *piano*. Although the language of the text is likely to limit the work's full appreciation, it is beautifully composed, and its density repays detailed study.

This density is likewise a feature of the orchestral Variations dedicated to the memory of Aldous Huxley, even though the first variation, after the opening cadences, is a monody shared between different groups of instruments – a very Stravinskian interpretation of Schoenberg's *Klangfarbenmelodie*. Here, as in all Stravinsky's later music, timbre and texture play a crucial role in articulating the form; the score contains his most extreme example of sustained polyrhythm – three sections for 12 rhythmically distinct instruments (cf the middle section of Messiaen's *Chronochromie*),

20. *Autograph MS of p.6 from Stravinsky's Variations for Orchestra (in memory of Aldous Huxley), 1963–4*

first violins, then violas with two double basses, and lastly wind. The complexity of these sections is the most difficult feature of the score to grasp; Stravinsky himself claimed that with repeated hearings they would appear to change perspective, like mobiles, but it is unfortunate that most hearings so far have been of the same (recorded) performance.

The last two in the series of sacred choral works that bulks so large in Stravinsky's final period, the *Introitus* and the *Requiem Canticles*, are far less hermetic in style. The former picks up and briefly reworks ideas from the Voice of God sequences in *The Flood*, with its two-part texture of muffled drumming (the pitches clarified by string doublings) and its solemn, unmistakably Stravinskian cadences. The *Requiem Canticles* are more complex and varied, but the textures are still relatively speaking simple, though the sonorities are used with unfailing imagination. The text, like that of the *Introitus*, is taken from the mass for the dead: part of the opening gradual, parts of the sequence 'Dies irae', and most of the final responsory 'Libera me'. They are prefaced by a prelude for strings alone (whose constantly repeated semiquavers provide a harmonic underpinning to the cumulatively polyrhythmic refrain above), separated by a woodwind interlude which Stravinsky himself referred to as 'the formal lament', and rounded off by a litany in which still chords for high wind, piano, harp and horn punctuate quicker moving chords on celesta, bells and vibraphone. Unlike *Abraham and Isaac* the *Requiem Canticles* do not eschew illustration of the text: the 'Dies irae' in particular is as vivid in its own rarefied context as Verdi's is in its. This 'pocket Requiem', as Stravinsky referred to it, is a distillation both of the liturgical text and of his own musical means of setting it, evolved and refined through a career of more than 60 years.

The conventional division of Stravinsky's mature music into three main periods – Russian, neo-classical and serial – has obvious justification, and even within these periods there are comparable if smaller-scale disjunctions of style. Equally obvious, at least to a generation familiar with the whole span of Stravinsky's oeuvre, and aware of analogous features in the work of Picasso and James Joyce, to mention only two of his greatest contemporaries in other media, is the music's underlying homogeneity, which makes almost any work from any period after 1910 immediately recognizable as his.

However, it is one thing to recognize this instinctively and quite another to define it in technically precise terms. This challenge has indeed only begun to be met within the last decade, though special mention must be made of Igor Glebov's (Boris Asaf'yev's) *Book on Stravinsky*, originally published in 1929 and therefore dealing mainly with the Russian period. Until the sketch-material in Stravinsky's personal archive becomes readily accessible, it will be impossible to give any comprehensive account of his compositional methods and preoccupations; he himself was always averse to giving any but the most oblique hints about them, and even these could be misleading (as Taruskin, 1980, shows as regards the folk material in *The Rite of Spring*).

What seems quite clear, however, is that Stravinsky was not interested in, and perhaps even rejected as a threat to spontaneity, any kind of unified-field theory such as has attracted many composers in Schoenberg's wake. Rather than attempt to create a new heaven and a new earth, Stravinsky preferred to take the existing musical world (defined in the widest historical terms) as his starting-point, and to impress his personality upon it: as he himself proclaimed, he was an evolutionary not a revolutionary composer. Any 'analysis' of his language which ignores or suppresses this contrapuntal relationship with the past risks being no more than descriptive; it can suggest the nature of the compositional choices he made, but not their rationale. It is for this reason that the deliberately ahistorical approach adopted by Forte (1978; see also Taruskin, 1979), though providing much raw material for a genuine understanding of *The Rite of Spring*, at times seems arbitrary and even irrelevant, obscuring rather than revealing the work's true coherence, which is based on an expansion of traditional (more specifically, Russian-traditional) tonal practice.

More successful, because more empirical and more ready to meet the music within its own context, is Van den Toorn's highly technical examination of a representative selection of works (1983). While making use of Forte's descriptive methods, this takes as its principal starting-point Arthur Berger's important article on 'Problems of Pitch Organization in Stravinsky' (in Boretz and Cone, 1968, 2/1972) and its recognition of the continuing importance in Stravinsky's vocabulary of the octatonic scale (alternating tones and semitones). More or less explicit references to this scale are found throughout Stravinsky's first two periods, and to some extent even in his serial music; they are discovered (in different ways at different times) to account both for the perceived identity of Stravinsky's style and for its characteristically ambiguous relationship to traditional tonality, which it simultaneously expands and subverts.

187

Indeed Stravinsky's use of the octatonic scale is in many ways analogous to Debussy's of the equally 'artificial' but more limited whole-tone scale, including his willingness to use it in impure forms.

Further characteristic elements of Stravinsky's language – rhythmic and structural as well as melodic and harmonic – have been identified and explored by various writers. Of these, Edward Cone's elucidations (1962 and 1968) of the structure of the Symphony in C, the Symphonies of Wind Instruments and other works seem to offer the greatest potentiality for further development, particularly as regards the tonal music (see Lang, 1963; Boretz and Cone, 1968, 2/1972). The serial music has found distinguished explicators in Milton Babbitt and Claudio Spies, perhaps because its strictness of organization relates more closely to their own compositional concerns. The non-professional listener should be warned, however, that much of the recent writing about Stravinsky's music is far harder to ingest than the music itself. The main, perhaps the only, constructive function of musical analysis is to clarify the implicit frame of reference in which the piece in question is meant to be heard, and in so doing to intensify the listener's appreciation of whatever it has to offer – good or bad, strong or weak, coherent or the reverse. If curiosity will not quite allow us to accept Stravinsky's own marked distaste for analysis as the last word on his music, it would at least be as well to recognize that its prime impact at all periods was physical, on the ear and the nervous system; cerebration can refine that response, but is no substitute for it.

WORKS

Publishers:

Associated [A]	Charling [Char]
Belyayev [Bel]	Chester [C]
Bessell [Bes]	Faber [F]
Boosey & Hawkes [B]	Hansen [H]
Breitkopf & Härtel [Br]	Henn [He]
Chappell [Chap]	Jurgenson [J]

Leeds [L]
Mercury [M]
Edition Russe de Musique [R]
Schott [S]
Sirène [Si]

Numbers in the right-hand column denote references in the text.

DRAMATIC

Title	Genre (acts, libretto/scenario)	Scoring	Composition	First performance	Publication	
Zhar'-ptitsa (L'oiseau de feu) [The firebird]	fairy story ballet (2 scenes, M. Fokin)	orch	1909–10	cond. G. Pierné, Paris, Opéra, 25 June 1910	J 1910, S	109, 111, 112–17, 118, 119, 128, 132, 134, 147
Petrushka (Pétrouchka)	burlesque (4 scenes, A. Benois)	orch	1910–11, rev. 1946	cond. P. Monteux, Paris, Châtelet, 13 June 1911	R 1912, rev. B 1947	113, 119, 124–5, 126, 131, 138, 142, 152, 160, 161
Vesna svyashchennaya (Le sacre du printemps) [The rite of spring (literally 'Sacred spring')]	scenes of pagan Russia (2 pts., N. Roerich)	orch	1911–13, Sacrificial Dance, rev. 1943	cond. Monteux, Paris, Champs-Elysées, 29 May 1913	R 1913 (for pf 4 hands), R 1921 (full score), rev. Sacrificial Dance A 1945, facs. sketches B 1969	54, 118, 119, 121, 125–8, *127*, 129, 136, 138, 141, 142, 145, 186, 187
Solovey (Le rossignol) [The nightingale]	musical fairy tale (3, Stravinsky, S. Mitusov after H. Andersen)	solo vv, chorus, orch	Act 1, 1908–9; Acts 2–3, 1913–14, rev.	cond. Monteux, Paris, Opéra, 26 May 1914	R 1923, B 1947, rev. B 1962	109, 123, 128–9, 134, 136, 161
Bayka (Renard) [Reynard]	burlesque in song and dance (Stravinsky after Russ. trad.)	2 T, 2 B, small orch	1915–16	Paris, Opéra, 18 May 1922	He 1917, C	131, 132, 133, 139, *140*, 141, 142, 147, 179, 183
Pesnya solov'ya (Chant du rossignol) [The song of the nightingale]	sym. poem/ballet (3 pts., Stravinsky after Andersen) [arr. from The nightingale]	orch	1917	concert perf. cond. E. Ansermet, Geneva, 6 Dec 1919; staged cond. Ansermet, Paris, Opéra, 2 Feb 1920	R 1921, B	134
Histoire du soldat (The Soldier's Tale)	to be read, played and danced (2 pts., C. F. Ramuz)	3 actors, female dancer, cl, bn, cornet, trbn, perc, vn, db	1918	cond. Ansermet, Lausanne, Municipal, 28 Sept 1918	C 1924	132, 134, 135, 142, 147, 167, 244, 252
Pulcinella	ballet with song (1) [after Pergolesi and others]	S, T, B, chamber orch	1919–20	cond. Ansermet, Paris, Opéra, 15 May 1920	C 1920 (vocal score), R (full score), B	135, 143, 147, 152

189

Title	Genre (acts, libretto/scenario)	Scoring	Composition	First performance	Publication	
Mavra	opera buffa (1, B. Kochno after Pushkin: The Little House in Kolomna)	S, Mez, A, T, orch	1921–2	cond. G. Fitelberg, Paris, Opéra, 3 June 1922	R 1925, B 1947	145, 147, 153–4
Svadebka (Les noces) [The Wedding]	Russ. choreographic scenes (4 scenes, Stravinsky after Russ. trad.)	S, Mez, T, B, SATB, 4 pf, perc ens	short score, 1914–17; scored, 1921–3	cond. Ansermet, Paris, Gaîté Lyrique, 13 June 1923	C 1922 (vocal score), C c1923 (full score)	124, 130–31, 132, 136, 137, 139–42, 147, 154, 179, 180, 183
2 frags. in abandoned scoring	opening section	solo vv, chorus, orch				
	scenes i–ii	solo vv, chorus, pianola, harmonium, 2 cimb, perc ens				
Oedipus rex	opera-oratorio (2, J. Cocteau after Sophocles, Lat. trans. J. Daniélou)	narrator, solo vv, male chorus, orch	1926–7, rev. 1948	concert perf. cond. Stravinsky, Paris, Sarah Bernhardt, 30 May 1927; staged Vienna, 23 Feb 1928	R 1927, rev. B 1949	147, 154, 155, 156, 161
Apollon musagète, rev. as Apollo	ballet (2 scenes)	str orch	1927–8, rev. 1947	Washington, DC, Library of Congress, 27 April 1928	R 1928, rev. B 1949	147, 154, 157, 161, 161, 174
Le baiser de la fée (The Fairy's Kiss)	allegorical ballet (4 scenes, Stravinsky after Andersen) [after songs and pf pieces by Tchaikovsky]	orch	1928, rev. 1950	cond. Stravinsky, Paris, Opéra, 27 Nov 1928	R 1928, rev. B 1952	148, 153, 154, 161
Perséphone (Persephone)	melodrama (3 scenes, A. Gide)	speaker, T, SATB, TrA, orch	1933–4, rev. 1949	cond. Stravinsky, Paris, Opéra, 30 April 1934	R 1934, rev. B 1950	150, 155, 156, 157, 160, 161, 166, 167
Jeu de cartes (The Card Party)	ballet in 3 deals (Stravinsky, M. Malaieff)	orch	1936	cond. Stravinsky, New York, Metropolitan, 27 April 1937	S 1937	151, 157
Circus Polka (for a young elephant)		band (scored D. Reksin)	1942	New York, Madison Square Gardens, spr. 1942	unpubd in this version	159, 164

Title	Scoring	Composition	First performance	Publication	
Scènes de ballet	for revue The Seven Lively Arts	orch	1944	cond. M. Abravanel, Philadelphia, 1944	Chap 1945	160, 164
Orpheus	ballet (3 scenes)	orch	1947	New York, City Center, 28 April 1948	B 1948	161, 165, 174, 178
The Rake's Progress	opera (3, W. H. Auden, C. Kallman)	solo vv, chorus, orch	1948–51	cond. Stravinsky, Venice, La Fenice, 11 Sept 1951	B 1951	162–3, 164, 166, 167, 168, 169, 173, 177
Agon	ballet	orch	1953–4, 1956–7	concert perf. cond. R. Craft, Los Angeles, 17 June 1957; staged, New York, 1 Dec 1957	B 1957	170–71, 174, 180, 181
The Flood	musical play (Craft after York and Chester mystery plays and Genesis)	T, 2B, SAT, actors, narrator, orch	1961–2	CBS television, broadcast 14 June 1962; staged cond. Craft, Hamburg, Staatsoper, 30 April 1963	B 1963	174, 182–3, 185

ORCHESTRAL

Title	Scoring	Composition	First performance	Publication	
Symphony no.1, Eb, op.1	orch	1905–7	cond. F. Blumenfeld, St Petersburg, 22 Jan 1908	J 1914	109, 113, 132
Scherzo fantastique, op.3	orch	1907–8	cond. A. Ziloti, St Petersburg, 6 Feb 1909	J, S	109, 115
Feu d'artifice (Fireworks), op.4	orch	1908	cond. A. Ziloti, St Petersburg, 6 Feb 1909	S 1910	109, 115, 134
Chant funèbre, op.5, lost	wind	1908	cond. Blumenfeld, St Petersburg, aut. 1908	unpubd	
Suite from 'The Firebird'	orch	1911		J 1912	
suite no.2	reduced orch	1919		C	
suite no.3	reduced orch	1945		L 1946–7	176
March [arr. of 3 Easy Pieces, pf 4 hands: no.1]	12 insts	1915		unpubd	
Rag-time	fl, cl, hn, cornet, trbn, perc, cimb, 2 vn, va, db	1918	cond. A. Bliss, London, 27 April 1920	Si 1919, C 1920	132, 133, 142
Symphonies of Wind Instruments	23 insts	1920, rev. 1945–7	cond. S. Koussevitzky, London, 10 June 1921	R 1926 (pf reduction), rev. B 1952	144, 152, 153, 161, 168, 180, 188
Suite no.2 [arr. of 3 Easy Pieces, pf 4 hands, and 5 Easy Pieces, pf 4 hands: no.5]	small orch	1921		C	

Title	Scoring	Composition	First performance	Publication	
Suite from 'Pulcinella'	chamber orch	c1922, rev. 1947	cond. Monteux, Boston, 22 Dec 1922	R 1924, rev. B 1949	161, 176
Concerto	pf, wind, timp, dbs	1923–4, rev. 1950	cond. Koussevitzky, Paris, Opéra, 22 May 1924	R 1924 (2 pf reduction), R 1936 (full score), rev. B C	59, 146, 154, 161
Suite no.1 [arr. of 5 Easy Pieces, pf 4 hands: nos.1–4]	small orch	1917–25			
Four Studies [arr. of 3 Pieces, str qt, and Study, pianola]	orch	nos.1–3, 1914–18; no.4, 1928	Berlin, 7 Nov 1930	R, B	
Capriccio	pf, orch	1928–9, rev. 1949	Stravinsky, cond. Ansermet, Paris, 6 Dec 1929	R 1930, rev. B 1952	146, 161, 168
Violin Concerto, D		1931	S. Dushkin, cond. Stravinsky, Berlin, 23 Oct 1931	S 1931	150
Divertimento [arr. from The Fairy's Kiss]	orch	1934, rev. 1949		R 1938, rev. B 1950	161
Preludium	jazz band	1936–7, reorchd 1953	rev. version, cond. Craft, Los Angeles, 18 Oct 1953	rev. B 1968	
Concerto 'Dumbarton Oaks', E♭	chamber orch	1937–8	cond. N. Boulanger, Washington, DC, 8 May 1938	S 1938	151, 157, 179
Symphony, C	orch	1938–40	cond. Stravinsky, Chicago, 7 Nov 1940	S 1948	151, 156, 158, 159, 163–4, 165, 168, 188
Danses concertantes	chamber orch	1941–2	cond. Stravinsky, Los Angeles, 8 Feb 1942	A 1942	159, 168
Circus Polka	orch	1942	cond. Stravinsky, Cambridge, Mass., 13 Jan 1944	A 1944	
Four Norwegian Moods	orch	1942	cond. Stravinsky, Cambridge, Mass., 13 Jan 1944	A 1944	159
Ode, elegiacal chant in 3 parts	orch	1943	cond. Koussevitzky, Boston, 8 Oct 1943	S 1947	159
Scherzo à la russe	jazz band	1943–4	cond. Stravinsky, San Francisco, March 1946	unpubd	159, 164
Symphony in Three Movements	orch	1942–5	cond. Stravinsky, New York, 24 Jan 1946	Chap 1945	159, 160, 165, 166
Ebony Concerto	cl, jazz band	1945	W. Herman, cond. W. Hendl, New York, 25 March 1946	A 1946	143, 154
Concerto, D	str	1946	cond. P. Sacher, Basle, 27 Jan 1947	Char 1946 B 1947	
Concertino [arr. of str qt work]	vn, vc, fl, ob, eng hn, A-cl, 2 bn, 2 tpt, trbn, b trbn	1952	Los Angeles, 11 Nov 1952	H 1953	161

Title, genre	Scoring	Composition	First performance	Publication	
Tango [arr. of pf work]	19 insts	1953	cond. Craft, Los Angeles, 18 Oct 1953	M 1954	
Greeting Prelude [after C. F. Summy: Happy Birthday to you]	orch	1955	cond. C. Munch, Boston, 4 April 1955	B 1956	172, 181, 182
Movements	pf, orch	1958–9	M. Weber, cond. Stravinsky, New York, 10 Jan 1960	B 1960	
Eight Instrumental Miniatures [arr. of Les cinq doigts, pf]	15 insts	1962	nos.1–4 cond. Craft, Los Angeles, 26 March 1962; nos.1–8 cond. Stravinsky, Toronto, 29 April 1962	C 1963	
Variations	orch	1963–4	cond. Craft, Chicago, 17 April 1965	B 1965	175, 183, 184
Canon on a Russian Popular Tune [theme from finale of The Firebird]	orch	1965	cond. Craft, Toronto, 16 Dec 1965	B 1966	

CHORAL

Title, genre	Text	Scoring	Composition	First performance	Publication	
cantata, lost			1904			
Zvezdolikiy (Le roi des étoiles) [The king of the stars (literally 'Star-faced')], cantata	Bal'mont	chorus, pf TTBB, orch	1911–12	cond. F. André, Brussels, 19 April 1939	J 1913	121, 126
Podblyudniya [Saucers] (Four Russian Peasant Songs)	after Russ. trad.	female vv, rev. for equal vv, 4 hn	1914–17, rev. 1954	cond. V. Kibalchich, Geneva, 1917	S c1930, C 1932, rev. C 1958	
1. U spasa v' Chigisakh [On saints' days in Chigisakh]		4vv	1916			
2. Ovsen' [Ovsen]		2vv	1917			
3. Shchuka [The pike]		3 solo vv, 4vv	1914			
4. Puzishche [Master Portly]		solo v, 4vv	1915			
Otche nash' [Our Father] rev. as Pater noster	Slavonic Lat.	SATB SATB	1926 1949		R 1932 B	148
Symphony of Psalms	Pss xxxviii.13–14, xxxix.2–4, cl	TrATB, orch	1930, rev. 1948	cond. Ansermet, Brussels, 13 Dec 1930	R 1930 (vocal score), R 1932 (full score), rev. B 1948	148, 155, 156, 161
Simvol' veri [Symbol of faith] rev. as Credo	Slavonic Lat.	SATB SATB	1932, rev. 1964 1949		R 1933 B	148

Title, genre	Text	Scoring	Composition	First performance	Publication	
Bogoroditse devo [Blessed Virgin] rev. as Ave Maria	Slavonic / Lat.	SATB / SATB	1934 / 1949		R 1934 / B	148
Babel, cantata	Genesis xi.1–9	male narrator, male vv, orch	1944	cond. W. Janssen, Los Angeles, 18 Nov 1945	S 1953	159
Mass	Lat.	TrATB, 2 ob, eng hn, 2 bn, 2 tpt, 3 trbn	1944–8	cond. Ansermet, Milan, 27 Oct 1948	B 1948	161–2, 165
Cantata	late medieval Eng. verse	S, T, female vv, 2 fl, ob, ob + eng hn, vc	1951–2	cond. Stravinsky, Los Angeles, 11 Nov 1952	B 1952	163, 178
Canticum sacrum ad honorem Sancti Marci nominis	Lat.	T, Bar, chorus, orch	1955	cond. Stravinsky, Venice, 13 Sept 1956	B 1956	170, 180, 182
Threni: id est Lamentationes Jeremiae prophetae	Lat.	S, A, 2T, 2B, chorus, orch	1957–8	cond. Stravinsky, Venice, 23 Sept 1958	B 1958	172, 180, 181, 182
A Sermon, a Narrative and a Prayer, cantata	St Paul, Acts, Dekker	A, T, speaker, chorus, orch	1960–61	cond. Sacher, Basle, 23 Feb 1962	B 1961	173, 174, 182
Anthem 'The Dove Descending Breaks the Air'	Eliot: Little Gidding, pt.IV	SATB	1962	cond. Craft, Los Angeles, 19 Feb 1962	appx to Expositions and Developments, 1962, B	
Introitus	Requiem	male vv, pf, harp, 2 timp, 2 tam-tams, va, dbs	1965	Chicago, 17 April 1965	B 1965	176, 185
Requiem Canticles		A, B, chorus, orch	1965–6		B 1967	176, 185

SOLO VOCAL

Storm Cloud (Pushkin), romance, 1v, pf, 1902, unpubd

The Mushrooms Going to War, song, B, pf, 1904, unpubd

Favn'i pastushka (Faune et bergère)[Faun and shepherdess], op.2, song suite, Mez, orch, 1906 (Bel 1908, B):   114
  1 Pastushka, 2 Favn', 3 Reka [Torrent]

Pastorale (textless), S, pf, 1907 (J 1910, C, S); arr. S, ob, eng hn, cl, bn, 1923 (S)

Deux mélodies, op.6 (Gorodetsky), Mez, pf, 1908 (J ?1912, B 1968):   114
  1 Vesna (Monastirskaya) [Spring (The cloister)]
  2 Rosyanka (Khlistorskaya) [A song of the dew (Mystic song of the ancient Russian flagellants)]

Deux poèmes de Paul Verlaine, op.9, Bar, pf, 1910 (J 1911, B 1954); arr. Bar, chamber orch, 1951 (B 1953):   125, 128, 150
  1 Un grand sommeil noir, 2 La lune blanche

Two Poems of Konstantin Bal'mont, S/T, pf, 1911 (R 1912, B); arr. S/T, 2 fl, 2 cl, pf, str qt, 1954 (B):   125, 126
  1 Nezabudoochka–tsvetochek' [The flower], 2 Golub' [The dove]

Tri stikhotvoreniya iz yaponskoy liriki (Trois poésies de la lyrique japonaise) [Three Japanese lyrics] (trans. A. Brandta), S, pf/(2 fl, 2 cl, pf, str qt), 1912–13 (R 1913, B):   121, 129
1 Akahito, 1912, 2 Mazatsumi, 1912, Tsaraiuki, 1913

Tri pesenki 'Iz' vospominaniy yunosheskikh' godov" [Three little songs 'Recollections of my childhood'] (Russ. trad.), 1v, pf, c1906, rev. 1913 (R 1914, B); arr. 1v, small orch, 1929–30 (R 1934, B):
1 Sorochen'ka [The magpie], 2 Vorona [The rook], 3 Chicher'yacher' [The jackdaw]

Pribaoutki (Russ. trad.), male v, fl, ob + eng hn, cl, bn, vn, va, vc, db, 1914 (He 1917, C):   131, 138
1 Kornilo [Kornillo], 2 Natashka, 3 Polkovnik' [The colonel], 4 Starets' i zayats' [The old man and the hare]

Berceuses du chat (Cat's Cradle Songs) (Russ. trad.), A, Eb-cl + cl, cl + A-cl, A-cl + b cl, 1915–16 (He 1917, C):   131, 138–9
1 Spi kot' [The tom-cat], 2 Kot' na pechi [The tom-cat on the stove], 3 Bay-bay [Bye-bye], 4 U kota kota [O tom-cat, tom-cat]

Trois histoires pour enfants (Russ. trad.), 1v, pf, 1915–17 (C 1920); no.1 arr. 1v, orch, 1923 (S):
1 Tilim'-bom' [Tilimbom], 1917, 2 Gusi, lebedi [Geese, swans], 1917, 3 Pesenka medvedya [The bear's little song], 1915

Berceuse (Stravinsky), 1v, pf, 1917 (in Eng. edn. of Expositions and Developments, 1962)

Four Russian Songs (Russ. trad.), 1v, pf, 1918–19 (C 1920):
1 Selezen (Khorovodnaya) [The drake (Round)], 1918
2 Zapevnaya [Counting-song], 1919
3 Podblyudnaya [Table-mat song], 1919
4 Sektantskaya [Dissident song], 1919

Chanson de Paracha [from Mavra], S, orch, 1922–3 (R ?1933)

Petit Ramusianum harmonique (Stravinsky), 1v/unison vv, 1937 (in Hommage à C.-F. Ramuz, Lausanne, 1938)

Petit canon pour la fête de Nadia Boulanger (J. de Meung), 2T, 1947, unpubd

Three Songs from William Shakespeare, Mez, fl, cl, va, 1953 (B 1954):   179
1 Musick to Heare, 2 Full Fathom Five, 3 When Daisies Pied

Four Songs (Stravinsky) [arrs. of 4 Russian Songs: nos. 1 and 4, and 3 histoires pour enfants: nos.2 and 1], 1v, fl, harp, gui, 1953–4 (C 1955):
1 The Drake, 1953, 2 A Russian Spiritual, 1954, 3 Geese and Swans, 1954, 4 Tilimbom, 1954

In memoriam Dylan Thomas (Thomas: Do not go gentle), dirge canons and song, T, str qt, 4 trbn, 1954 (B 1954)   169, 170, 175, 179

Abraham and Isaac (Genesis xxii, in Heb.), sacred ballad, Bar, chamber orch, 1962–3 (B 1965)   175, 183, 185

Elegy for J. F. K. (Auden), Bar, 3 cl, 1964 (B 1964); rev. Mez, 3 cl, 1964 (B 1964)   175

The Owl and the Pussy-cat (Lear), 1v, pf, 1966 (B 1967)   176

CHAMBER AND INSTRUMENTAL

Three Pieces, str qt, 1914 (R 1922, B 1947)   124, 136, 152
Polka [arr. of 3 Easy Pieces, pf 4 hands: no.3], cimb, 1915, unpubd
Canons, 2 hn, 1917, unpubd
Study, pianola, 1917, roll 1967 B (Aeolian Co.), unpubd in score   134, 142
Duet, 2 bn, 1918, unpubd
Suite from 'The Soldier's Tale', cl, bn, cornet, trbn, perc, vn, db, 1918 (C 1922); 5 movts arr. vn, cl, pf, 1919 (C 1920)
Three Pieces, cl + A-cl, 1919 (C 1920)
Concertino, str qt, 1920 (H 1923)   152
Octet, fl, cl, 2 bn, C-tpt, A-tpt, trbn, b trbn, 1922–3 (R 1924), rev. 1952 (B 1952)   145, 151, 154, 155, 161, 178
Suite d'après thèmes, fragments et pièces de Giambattista Pergolesi [arr. from Pulcinella], vn, pf, 1925 (R 1926, B)
Prélude et Ronde des princesses [arr. from The Firebird], vn, pf, 1929 (S); Ronde rev. as Scherzo, 1933, collab. Dushkin (S)
Berceuse [arr. from The Firebird], vn, pf, 1929 (S); rev. 1933, collab. Dushkin (S)
Duo concertant, vn, pf, 1931–2 (R 1933, B)   150, 155
Chants du rossignol et Marche chinoise [arr. from The Nightingale], vn, pf, 1932, collab. Dushkin (R, B)

Suite italienne [arr. from Pulcinella], vc, pf, 1932, collab. Piatigorsky (R 1934, B)
Suite italienne [arr. from Pulcinella], vn, pf, 1932, collab. Dushkin (R 1934, B)

Divertimento [arr. of orch work], vn, pf, 1932, collab. Dushkin (R, B)
Pastorale [arr. of vocalise], vn, pf, 1933, collab. Dushkin (S 1934); arr. vn, ob, eng hn, cl, bn, 1933 (S 1934)
Chanson russe [arr. from Mavra], vn, pf, 1937, collab. Dushkin (R 1938, B); arr. vc, pf, collab. D. Markevich (R, B)
Élégie, va, 1944 (Chap 1945)
Ballad [arr. from The Fairy's Kiss], vn, pf, 1947, collab. J. Gautier (B 1951)

Septet, cl, bn, hn, pf, vn, va, vc, 1952–3 (B 1953)   179
Epitaphium, fl, cl, harp, 1959 (B 1959)   175
Double Canon, str qt, 1959 (B 1960)   175
Lullaby [arr. from The Rake's Progress], tr rec, a rec (B 1960)
Fanfare for a New Theatre, 2 tpt, 1964 (B 1968)

PIANO

Tarantella, 1898, unpubd
Scherzo, 1902 (F 1975)   112
Sonata, f♯, 1903–4 (F 1975)   108, *110*, 112–13, 114–15

Four Studies, op.7, 1908 (J 1910)
Valse des fleurs, 2 pf, 1914, unpubd, lost
Three Easy Pieces, 4 hands, 1914–15 (He 1917, C):   142
  1 March, 2 Waltz, 3 Polka
Souvenir d'une marche boche, 1915 (in E. Wharton, ed.: *The Book of the Homeless*, London, 1916)
Five Easy Pieces, 4 hands, 1916–17 (He 1917, C):   134, 142
  1 Andante, 2 Española, 3 Balalaika, 4 Napolitana, 5 Galop
Valse pour les enfants, c1917 (in *Le figaro*, 21 May 1922)   142
Piano-rag-music, 1919 (C 1920)
Les cinq doigts, 8 easy pieces, 1921 (C 1922)
Three Movements from 'Petrushka', 1921 (R 1922, B)   146

Sonata, 1924 (R 1925, B)   147, 155
Serenade, A, 1925 (R 1926, B)   147, 155
Concerto, 2 pf, 1931–5 (S 1936)   146, 151, 157, 178

Berceuses à 2 mains, 1940, unpubd
Tango, 1940 (M 1941)
Sonata, 2 pf, 1943–4 (Chap 1945)

REDUCTIONS OF OWN WORKS

Arrangements intended as independent works are listed above; the following reductions were made only for rehearsal or amateur use.
Vocal scores: Faun and Shepherdess, The King of the Stars, The Nightingale, Pribaoutki, Cat's Cradle Songs, Reynard, Pulcinella, Mavra, The Wedding, Oedipus rex, Babel, Cantata, Three Songs from William Shakespeare, In memoriam Dylan Thomas, Canticum sacrum
Pf solo: The Firebird, The Song of the Nightingale, The Soldier's Tale, Rag-time, Apollon musagète, The Fairy's Kiss, The Card Party, Preludium, Circus Polka
Pf 4 hands: Petrushka, The Rite of Spring, Concertino
2 pf: Concerto for pf and wind, Capriccio, Concerto 'Dumbarton Oaks', Septet, Agon, Movements
Vn, pf: Violin Concerto

ARRANGEMENTS

E. Grieg: Kobold, orch (for ballet Le festin), 1909, unpubd
F. Chopin: Nocturne, A♭; Valse brillante, E♭, orch, 1909 (for ballet)   111
Two Songs of the Flea (Goethe) [arrs. of Beethoven: op.75 no.3 and Musorgsky], B, orch, 1910 (Bes, B)   111
M. Musorgsky: Khovanshchina, 1913, collab. Ravel; unpubd except for vocal score of Stravinsky's final chorus, based on theme by Musorgsky (Bes 1914)   121
Song of the Volga Boatmen, orch, 1917 (C 1920)
M. Musorgsky: Boris Godunov: Prologue, pf, 1918, unpubd   134

R. de Lisle: La marseillaise, vn, 1919, unpubd

P. Tchaikovsky: The Sleeping Beauty: Variation d'Aurore; Entr'acte symphonique, orch, 1921, unpubd; Bluebird Pas-de-deux, small orch, 1941 (S 1953) — 145

The Star-spangled Banner, orch, 1941 (M)

J. S. Bach: Choral-Variationen über das Weihnachtslied 'Vom Himmel hoch da komm' ich her', chorus, orch, 1955-6 (B 1956) — 170

C. Gesualdo di Venosa: Tres sacrae cantiones, sextus and bassus parts supplied, 1957-9 (B 1957 [no.3], B 1960 [complete]): 1 Da pacem Domine, 1959, 2 Assumpta est Maria, 1959, 3 Illumina nos, 1957

Monumentum pro Gesualdo di Venosa ad CD annum [arrs. of madrigals Asciugate i begli occhi, Ma tu, cagion di quella and Belta poi che t'assenti], orch, 1960 (B 1960)

J. Sibelius: Canzonetta, op.62a, 2 cl, 4 hn, harp, db, 1963 (Br 1964)

H. Wolf: Two Sacred Songs [from the Spanisches Liederbuch], Mez, 9 insts, 1968 (B) — 176

J. S. Bach: Two Preludes and Fugues [from the '48'], str, ww, c1969, unpubd — 176

WRITINGS

with W. Nouvel: Chroniques de ma vie (Paris, 1935-6, 2/1962; Eng. trans., 1936; Eng. trans. as An Autobiography, 1936/R1975; Sp. trans., 1936-7; Ger. trans., 1937; Russ. trans., 1964; Bulg. trans., 1966; Hung. trans., 1969) — 151

Poétique musicale (Cambridge, Mass., 1942; Eng. trans., 1947; Ger. trans., 1949, 3/1966; It. trans., 1954; Dan. trans., 1961; Rom. trans., 1967; Eng.-Fr. edn., 1970) — 158

Leben und Werk (Zurich and Mainz, 1957) [reprints of Ger. trans. of Chroniques de ma vie and Poétique musicale with 'Answers to 35 Questions', incl. one question omitted in Conversations] — 171

with R. Craft: Conversations with Igor Stravinsky (London and New York, 1959; Russ. trans., 1971)

———: Memories and Commentaries (London and New York, 1960; Russ. trans., 1971) — 171

———: Dialogues and a Diary (New York, 1961, enlarged London, 1968; Russ. trans., 1971) — 171

———: Stravinsky in Conversation with Robert Craft (Harmondsworth, 1962; Ger. trans., 1961) [= Conversations and Memories]

———: Expositions and Developments (London and New York, 1962; Russ. trans., 1971) — 171

———: Themes and Episodes (New York, 1966, 2/1967) — 171

———: Retrospectives and Conclusions (New York, 1969) — 171

———: Themes and Conclusions (London, 1972; Ger. trans., 1972) [Themes and Episodes and Retrospectives and Conclusions]

ed. L. Kutateladse: Stati, pisma, vospominaniya [Articles, letters, memoirs] (Leningrad, 1972)

ed. L. S. D'yachkova: Stati i materiali [Articles and materials] (Moscow, 1973) [incl. 60 letters]

BIBLIOGRAPHY

CATALOGUES AND BIBLIOGRAPHIES

P. Magriel: Bibliography, *Stravinsky in the Theatre*, ed. M. Lederman (London and New York, 1949/*R*1975)

Igor Stravinsky: a Catalogue of his Published Compositions (London, 1957, rev., 2/1962)

C. D. Wade: 'A Selected Bibliography of Igor Stravinsky', *MQ*, xlviii (1962), 372; repr. in *Stravinsky: a New Appraisal of his Work*, ed. P. H. Lang (New York, 1963), 97

S. Cohen: *Stravinsky and the Dance: a Survey of Ballet Productions 1910–1962* (New York, 1962)

Stravinsky and the Theatre: a Catalogue of Decor and Costume Designs for Stage Productions of his Works (New York, 1963)

E. W. White: *Stravinsky: the Composer and his Works* (London, 1966, rev., enlarged 2/1979) [pt.2: register of works; appx C: catalogue of MSS (1904–52) in Stravinsky's possession]

D. Hamilton: 'Igor Stravinsky: a Discography of the Composer's Performances', *Perspectives on Schoenberg and Stravinsky*, ed. B. Boretz and E. T. Cone (Princeton, NJ, 1968, 2/1972); repr. in *PNM*, ix/2–x/1 (1971), 163

I. Beletzky and I. Blazhkov: 'Spisok proizvedenii I. F. Stravinskovo' [A list of Stravinsky's works], in I. Stravinsky and R. Craft: *Dialogi* (Leningrad, 1971), 375

J. Heiss: 'Index to Stravinsky and Craft', *PNM*, ix/2–x/1 (1971), 159

R. Kraus: 'Bibliographie, Igor Strawinsky', *Musik und Bildung*, iii (1971), 304

C. Spies: 'Editions of Stravinsky's Music', *Perspectives on Schoenberg and Stravinsky*, ed. B. Boretz and E. T. Cone (New York, 2/1972), 250

Igor Strawinsky (1882–1971): Phonographie (Frankfurt am Main, 1972)

D.-R. de Lerma, ed.: *Igor Fedorovitch Stravinsky: a Practical Guide to Publications of his Music* (Kent, Ohio, 1974)

K. Bailey: 'The Craft/Stravinsky Books: Bibliography and Commentary', *Studies in Music*, iii (1978), 48

C. Caesar: *Igor Stravinsky: a Complete Catalogue* (San Francisco, 1982)

C. M. Joseph: 'Stravinsky Manuscripts in the Library of Congress and the Pierpont Morgan Library', *Journal of Musicology*, i (1982), 327

PRIMARY MATERIALS

Avec Stravinsky: textes d'Igor Stravinsky et al (Monaco, 1958)

I. F. Stravinsky and R. Craft: *Conversations with Igor Stravinsky*

(London and New York, 1959; Russ. trans., 1971) [contains letters from Debussy, Ravel, Rivière, Dylan Thomas]

———: *Memories and Commentaries* (London and New York, 1960; Russ. trans., 1971) [contains letters from Auden, Benois, Lord Berners, Dyagilev, Gide, Nouvel, Prokofiev]

V. V. Yastrebstev: *Moi vospominaniya o N. A. Rimskom-Korsakove*, ii (Leningrad, 1962) [Chronicle of Rimsky-Korsakov's life, 1902–8, with many references to Stravinsky]

E. W. White: *Stravinsky: the Composer and his Works* (London, 1966, rev., enlarged 2/1979) [appx B: selection of letters written to Stravinsky in 1913]

A. Newman: *Bravo Stravinsky!* (Cleveland, Ohio, 1967)

P. Meylan: 'Ernest Ansermet und Igor Stravinsky: unveröffentlichte Briefe zur Freundschaft und Entfremdung', *Neue Zürcher Zeitung*, xli (1970), 51

I. Beletzky: notes in I. Stravinsky and R. Craft: *Dialogi* (Leningrad, 1971) [from four 'Conversations' books]

E. W. White: 'Stravinsky in Interview', *Tempo*, no.97 (1971), 6

C. Spies: 'Impressions after an Exhibition', *Tempo*, no.102 (1972), 2

L. Kutateladze, ed.: *F(yodor) I(gnat'yevich) Stravinsky: Stat'i, pis'ma, vospominaniya* [Articles, letters, memoirs] (Leningrad, 1972)

T. Stravinsky: *Catherine and Igor Stravinsky: a Family Album* (London, 1973)

I. Blazhkov, ed.: 'Pis'ma I. F. Stravinskovo', *Stat'i i materialï*, compiled L. S. D'yachkova (Moscow, 1973), 435–520 [62 letters, 1906–45, with commentary]

V. Stravinsky and R. Craft: *Stravinsky in Pictures and Documents* (London and New York, 1978)

R. Craft: 'Dear Bob[sky] (Stravinsky's Letters to Robert Craft, 1944–49)', *MQ*, lxv (1979), 392–439

———: 'Stravinsky Pre-centenary', *PNM*, xix (1980–81), 464

———: 'The Singular Joy of Scrutinizing Stravinsky: Letters from Emile Jaques-Dalcroze', *Dance Magazine*, iv (1981), 80

———: *Igor and Vera Stravinsky: a Photograph Album 1921 to 1971* (London, 1982) [contains interviews and miscellaneous writings]

———: 'The Story Behind Stravinsky's Rejection by l'Institut de France', *Ovation*, iii (1982), 12, 33

———: 'Stravinsky: Letters to Pierre Boulez', *MT*, cxxiii (1982), 396

———: *Stravinsky: Selected Correspondence*, i (New York, 1982)

COLLECTIONS OF ESSAYS

ReM, v/2 (1923) [special issue]

Contemporaneos, v/15 (Mexico City, 1929) [special issue]

Cahiers de Belgique, iii/10 (1930) [special issue]

Neujahreblatt der Allgemeinen Musikgesellschaft in Zürich, no.121 (1933) [special issue]

M. Armitage, ed.: *Igor Strawinsky* (New York, 1936; rev., enlarged 2/1949, ed. E. Corle)

ReM, no.191 (1939) [special issue]

Dance Index, vi/10–12 (1947) [special issues]; repr. in *Stravinsky in the Theatre*, ed. M. Lederman (London and New York, 1949/*R*1975)

Tempo, no.8 (1948) [special issue]

E. Corle, ed.: *Igor Stravinsky* (New York, 1949) [rev. edn. of Armitage, 1936]

M. Lederman, ed.: *Stravinsky in the Theatre* (London and New York, 1949/*R*1975)

Musik der Zeit, ix (1952) [special issue]

The Score, no.20 (1957) [special issue]

M. Cosman and H. Keller: *Stravinsky at Rehearsal* (London, 1962; Ger. trans., 1962)

Feuilles musicales, xv/2–3 (1962) [special issue]

MQ, xlviii/3 (1962) [special issue]; repr. as *Stravinsky: a New Appraisal of his Work*, ed. P. H. Lang (New York, 1963)

Tempo, no.61–2 (1962) [special issue]

Tempo, no.81 (1967) [special issue]

B. Boretz and E. T. Cone, eds.: *Perspectives on Schoenberg and Stravinsky* (Princeton, NJ, 1968, 2/1972)

Stravinsky: génies et réalités (Paris, 1968) [includes catalogue and critical discography by Harry Halbreich]

'The Composer in Academia: Reflections on a Theme of Stravinsky', *College Music Symposium*, x (1970), 55–98

Melos, xxxviii/9 (1971) [special issue]

PNM, ix/2–x/1 (1971) [special issue]

Tempo, no.97 (1971) [special issue]

Les cahiers canadiens de musique, no.4 (1972) [special issue: *Dossier Stravinsky: Canada 1937–67*]

D. Revesz, ed.: *In memoriam Igor Stravinsky* (Budapest, 1972)

L. S. D'yachkova and B. M. Yarustovsky, eds.: *Stati i materiali* [Articles and materials] (Moscow, 1973)

N. Goldner: *The Stravinsky Festival of the New York City Ballet* (New York, 1973)

Ars, revista de arte, dedicada a Igor Stravinsky (Buenos Aires, 1979)

L. Goehr: *Stravinsky Rehearses Stravinsky* (London, 1982)

H. Keller and M. Cosman: *Stravinsky Seen and Heard* (London, 1982)

H. Lindlar, ed.: *Igor Strawinsky: Aufsaetze, Kritiken, Erinnerungen* (Frankfurt am Main, 1982)

MQ, lxviii/4 (1982) [special issue]

Bibliography

C. Oja, ed.: *Stravinsky in Modern Music* (New York, 1982) [articles repr. from *MM*, 1924–46]

A. Schouvaloff and V. Borovsky: *Stravinsky on Stage* (London, 1982)

Tempo, no.141 (1982) [special issue]

J. Pasler, ed.: *Stravinsky: Centennial Essays* (Berkeley, Calif., in preparation)

MONOGRAPHS

A. Casella: *Igor Strawinski* (Rome, 1926, enlarged 2/1947, 3/1961)

Yu. Vainkop: *Stravinsky* (Leningrad, 1927)

B. de Schloezer: *Igor Stravinsky* (Paris, 1929); Eng. trans. in *The Dial*, lxxxv/6 (1929), abridged in *Igor Stravinsky*, ed. E. Corle (New York, 1949), 33–91; see also P. Landormy, *Musique*, ii (1929), 933

I. Glebov [B. V. Asaf'yev]: *Kniga o Stravinskom* [Book on Stravinsky] (Leningrad, 1929, repr. 1977 with introduction by B. Yarustovsky; see also P. Souvtchinsky, *Musique*, iii/6 (1930); Eng. trans. as *A Book about Stravinsky* (Ann Arbor, 1982)

P. Collaer: *Strawinsky* (Brussels, 1930)

E. W. White: *Stravinsky's Sacrifice to Apollo* (London, 1930)

D. de'Paoli: *L'opera di Strawinsky* (Milan, 1931)

H. Fleischer: *Strawinsky* (Berlin, 1931)

A. Schaeffner: *Strawinsky* (Paris, 1931)

D. de'Paoli: *Igor Strawinsky: da 'L'oiseau de feu' a 'Persefone'* (Turin, 1934)

G. F. Malipiero: *Strawinsky* (Venice, 1945)

E. W. White: *Stravinsky: a Critical Survey* (London, 1947; Ger. trans., 1950)

F. Onnen: *Stravinsky* (Stockholm and London, 1948)

A. Tansman: *Igor Stravinsky* (Paris, 1948; Eng. trans., 1949; Sp. trans., 1949)

J. E. Cirlot: *Igor Strawinsky: su tiempo, su significación, su obra* (Barcelona, 1949)

R. H. Myers: *Introduction to the Music of Stravinsky* (London, 1950)

L. Oleggini: *Connaissance de Stravinsky* (Lausanne, 1952)

J. van Ackere: *Igor Strawinsky* (Antwerp, 1954)

H. Strobel: *Stravinsky: Classic Humanist* (New York, 1955/R1973; Ger. orig., 1956)

F. Sopeña: *Strawinsky: vida, obra y estilo* (Madrid, 1956)

H. H. Stuckenschmidt: *Strawinsky und sein Jahrhundert* (Berlin-Dahlem, 1957)

H. Kirchmeyer: *Igor Strawinsky: Zeitgeschichte im Persönlichkeitsbild* (Regensburg, 1958)

M. Monnikendam: *Strawinsky* (Haarlem, 1958)

R. Vlad: *Strawinsky* (Rome, 1958; Eng. trans., 1960, rev., enlarged 3/1979)

R. Siohan: *Stravinsky* (Paris, 1959, 2/1971; Ger. trans., 1960; Eng. trans, 1966)

F. Herzfeld: *Igor Strawinsky* (Berlin, 1961)

L. Fábián: *Igor Sztravinszkij* (Budapest, 1963)

B. Yarustovsky: *Igor' Stravinsky* (Moscow, 1963, 2/1969; enlarged 3/1982; Ger. trans., 1966)

N. Nabokov: *Igor Strawinsky* (Berlin, 1964)

G. Tintori: *Stravinsky* (Milan, 1964; Fr. trans., 1966)

G. Berger: *Igor Strawinsky* (Wolfenbüttel, 1965)

P. Faltin: *Igor Stravinsky* (Bratislava, 1965)

M. Philippot: *Igor Stravinsky* (Paris, 1965)

E. W. White: *Stravinsky: the Composer and his Works* (London, 1966, rev., enlarged 2/1979)

O. Nordvall: *Igor Stravinsky: ett porträtt med citat* (Stockholm, 1967)

P. M. Young: *Stravinsky* (New York and London, 1969)

A. Dobrin: *Igor Stravinsky* (New York, 1970)

R. M. Trahey: *The Aesthetics of Stravinsky's Musical Style: the Relationship of Culture and the Arts* (diss., U. of Eastern Illinois, 1972)

M. Druskin: *Igor' Stravinsky: lichnost', tvorchestvo, vzglyadï* [Personality, works, views] (Leningrad, 1974, rev., enlarged 2/1979; Ger. trans., 1976; Eng. trans., 1983)

F. Routh: *Stravinsky* (London, 1975/*R*1977)

I. F. Andras: *Igor Sztravinszkij* (Budapest, 1976)

N. Tierney: *The Unknown Country: a Life of Igor Stravinsky* (London, 1977)

L. Erhardt: *Igor Strawinski* (Warsaw, 1978)

K. McLeish: *Stravinsky* (London, 1978)

R. Nichols: *Stravinsky* (London and Milton Keynes, 1978)

W. Dömling: *Igor Strawinsky in Selbstzeugnissen und Bilddokumenten* (Reinbek bei Hamburg, 1982)

T. Hirsbrunner: *Igor Strawinsky in Paris* (Laaber, 1982)

P. C. Van den Toorn: *The Music of Igor Stravinsky* (New Haven and London, 1983)

MEMOIRS

C. F. Ramuz: *Souvenirs sur Igor Strawinsky* (Lausanne, 1929, 3/1952; Ger. trans., *c*1956; excerpts trans. in *Stravinsky in the Theatre*, ed. M. Lederman (London and New York, 1949/*R*1975)

L. Kerstein: 'Working with Stravinsky on *Jeu de Cartes*', *MM*, xiv/3

Bibliography

(1937); repr. in *Stravinsky in the Theatre*, ed. M. Lederman (London and New York, 1949/*R*1975), 136

F. Jacobi: 'Harvard Soirée', *MM*, xvii (1939), 47

G. Antheil: *Bad Boy of Music* (New York, 1945), 30ff

T. Karsavina: 'A Recollection of Stravinsky', *Tempo*, no.8 (1948), 7

S. Dushkin: 'Working with Stravinsky', *Igor Stravinsky*, ed. E. Corle (New York, 1949), 179

N. Nabokov: 'Christmas with Stravinsky', *Igor Stravinsky*, ed. E. Corle (New York, 1949), 123–68

H. Raynor: 'Stravinsky the Teacher', *The Chesterian*, xxxi (1956), 35, 69

R. Craft: 'A Personal Preface', *The Score*, no.20 (1957), 7

M. Perrin: 'Stravinsky in a Composition Class', *The Score*, no.20 (1957), 44

D. Godowsky: *First Person Plural* (New York, 1958)

R. Palmer: 'Stravinsky in Russia', *New Statesman and Nation* (2 Nov 1962), 613

A. Afonina: 'Igor' Stravinskii v sovetskom soyuze' [Igor Stravinsky in the USSR], *SovM* (1963), no.1, p.125

I. Raskin: 'Diary of a Recording Artist', *Music Journal*, xxiii/2 (1965), 33

M. Babbitt: [untitled memoir], *PNM*, ix/2–x/1 (1971), 87

E. Carter: [untitled memoir], *PNM*, ix /2–x/1 (1971), 1

B. Johnston: 'An Interview with Soulima Stravinsky', *PNM*, ix/2–x/1 (1971), 15

E. Krenek: [untitled memoir], *PNM*, ix/2–x/1 (1971), 7

W. Piston: 'A Reminiscence', *PNM*, ix/2–x/1 (1971), 6

L. Smit: 'A Card Game, a Wedding and a Passing', *PNM*, ix/2–x/1 (1971), 87

V. Ussachevsky: 'My Saint Stravinsky', *PNM*, ix/2–x/1 (1971), 34

R. Craft: *Stravinsky: Chronicle of a Friendship 1948–1971* (London and New York, 1972)

P. Horgan: *Encounters with Stravinsky: a Personal Record* (London and New York, 1972)

L. Libman: *And Music at the Close: Stravinsky's Last Years: a Personal Memoir* (New York, 1972)

T. Stravinsky: *Catherine and Igor Stravinsky* (London, 1973)

N. Nabokov: *Bagazh* (New York, 1975) [reminiscences]

K. Stravinskaya: *Ob I. F. Stravinskom i evo blizkikh* [Stravinsky and his intimates] (Leningrad, 1978)

S. Isacoff: 'Musical Life with Father', *Key Classics*, i/5 (1981), 16

A. Carpentier: 'Klassitsizm i galstuki' [Classicism and neckties], *SovM* (1982), no.7, 95

R. Craft: 'My Life with Stravinsky', *New York Review of Books* (10 June 1982), 6

——: 'On a Misunderstood Collaboration: Assisting Stravinsky', *Atlantic Monthly* (Dec 1982), 68

R. Kirkpatrick: 'Recollections of Two Composers: Hindemith and Stravinsky', *Yale Review*, lxxi (1982), 627

BALLETS RUSSES

W. A. Propert: *The Russian Ballet in Western Europe, 1909–1920* (London, 1921/*R*1972)

A. Haskell with W. Nouvel: *Diaghileff: his Artistic and Private Life* (New York, 1935/*R*1978)

S. Lifar: *Serge Diaghilev: his Life, his Work, his Legend* (New York, 1940)

S. L. Grigoriev: *The Diaghilev Ballet 1909–1929* (London, 1953)

B. Kochno: *Diaghilev and the Ballets Russes* (New York, 1970)

J. Percival: *The World of Diaghilev* (London, 1971)

C. Spencer: *The World of Serge Diaghilev* (London, 1974)

R. Buckle: *Diaghilev* (New York, 1979)

CRITICAL EVALUATIONS

V. Karatïgin: 'Molodïye russkiye kompozitorï' [Young Russian composers], *Apollon* (1910), no. 11, p.30

M. Calvocoressi: 'A Russian Composer of To-day', *MT*, lii (1911), 511

C. Van Vechten: 'Igor Stravinsky: a New Composer', *Music After the Great War* (New York, 1915), 85–117

C. S. Wise: 'Impressions of Igor Stravinsky', *MQ*, ii (1916), 249

R. D. Chennevière [D. Rudhyar]: 'The Two Trends of Modern Music in Stravinsky's Works', *MQ*, v (1919), 169

L. Henry: 'Igor Stravinsky and the Objective Direction in Contemporary Music', *The Chesterian*, 2nd ser., no.4 (1920), 97

J. Cocteau: 'Stravinsky, dernière heure', *ReM*, v/2 (1923), 31

A. Levinson: 'Stravinsky et la danse', *ReM*, v/2 (1923), 17

B. de Schloezer: 'Igor Strawinsky und Serge Prokofjeff', *Melos*, iv (1924), 469

P. Rosenfeld: 'Musical Chronicle', *The Dial*, lxxviii (1925), 259

N. Boulanger: *Lectures on Modern Music, Delivered under the Auspices of the Rice Institute Lectureship in Music* (Houston, 1926)

L. Sabaneyev: *Modern Russian Composers* (New York, 1927/ *R*1975), 64ff

A. Lourié: 'Neo-gothic and Neo-classic', *MM*, v/3 (1928), 3

L. Sabaneyev: 'The Stravinsky Legends', *MT*, lxix (1928), 785

——: 'Dawn or Dusk? Stravinsky's New Ballets, "Apollo" and "The Fairy's Kiss"', *MT*, lxx (1929), 403

Bibliography

P. Rosenfeld: 'European Music in Decay', *Scribner's Magazine*, lxxxix (1931), 277

———: 'The Two Stravinskys', *The New Republic*, lxvi (1931), 20

A. Al'shvang: 'Ideynïy put' Stravinskovo [Stravinsky's ideological path], *SovM* (1933), no.5, p.29; repr. in *Izbrannïye stat'i* (Moscow, 1959), 297

J. Handschin: *Igor Strawinski: Versuch einer Einführung* (Zurich, 1933)

C. Lambert: 'Diaghilev and Stravinsky as Time Travellers', *Music, Ho!* (London, 1934), 69ff

A. Schaeffner: 'On Stravinsky, Early and Late', *MM*, xii (1934), 3

M. Blitzstein: 'The Phenomenon of Stravinsky', *MQ*, xxi (1935), 330

A. Hoérée: 'Invention pure et matière musicale chez Strawinsky', *ReM*, no.191 (1939), 100

S. Lifar: 'Igor Strawinsky, législateur du ballet', *ReM*, no.191 (1939), 81

L. Saminsky: *Music of Our Day* (New York, 2/1939), 223–320

A. Schaeffner: 'Critique et thématique', *ReM*, no.191 (1939), 1

P. Souvtchinsky: 'La notion du temps et la musique', *ReM*, no.191 (1939), 70 [source for the discussion of time in *Poetics of Music*]

A. Kall: 'Stravinsky in the Chair of Poetry', *MQ*, xxvi (1940), 283

N. Nabokov: 'Stravinsky Now', *Partisan Review*, xi (1944), 324

P. Souvtchinsky: 'Igor Stravinsky', *Contrepoints*, ii (1946), 19

E. Ansermet: 'Stravinsky's Gift to the West', *Dance Index*, vi (1947), 235

A. Berger: 'Music for the Ballet', *Dance Index*, vi (1947), 258

R. Leibowitz: 'Schoenberg and Stravinsky', *Partisan Review*, xv (1948), 361

N. Nabokov: 'The Atonal Trail', *Partisan Review*, xv (1948), 580 [reply to Leibowitz, 1948]

———: 'Atonality and Obscurantism', *Partisan Review*, xv (1948), 1148

T. Strawinsky: *Le message d'Igor Strawinsky* (Lausanne, 1948; Ger. trans., 1952; Eng. trans., 1953)

T. W. Adorno: *Philosophie der neuen Musik* (Tübingen, 1949, 3/1969; Fr. trans., 1962; Hung. trans., 1970; Eng. trans., 1973)

G. Balanchine: 'The Dance Element in Stravinsky's Music', *Stravinsky in the Theatre*, ed. M. Lederman (London and New York, 1949/R1975), 75

A. Berger: 'The Stravinsky Panorama', *Igor Stravinsky*, ed. E. Corle (New York, 1949), 105

H. Boys: 'Organic Continuity', *Igor Stravinsky*, ed. E. Corle (New York, 1949), 93

———: 'Stravinsky: Critical Categories Needed for a Study of his Music', *The Score*, no.1 (1949), 3

A. Copland: 'The Personality of Stravinsky', *Igor Stravinsky*, ed. E. Corle (New York, 1949), 121

L. Morton: 'Incongruity and Faith', *Igor Stravinsky*, ed. E. Corle (New York, 1949), 193

N. Nabokov: 'Igor Stravinsky', *Atlantic Monthly*, clxxxiv (1949), 21

——: 'Stravinsky and the Drama', *Stravinsky in the Theatre*, ed. M. Lederman (London and New York, 1949/*R*1975), 104

E. Satie: 'A Composer's Conviction', *Igor Stravinsky*, ed. E. Corle (New York, 1949), 25

H. Boys: 'Stravinsky: à propos his Aesthetic', *The Score*, no.2 (1950), 61

——: 'Stravinsky: the Musical Materials', *The Score*, no.4 (1951), 31

H. Murrill: 'Aspects of Stravinsky', *ML*, xxxii (1951), 118

H. Keller: 'Schönberg and Stravinsky: Schönbergians and Stravinskians', *MR*, xv (1954), 307

A. Berger: 'Stravinsky and the Younger American Composer', *The Score*, no.12 (1955), 38

H. Boys: 'A Note on Stravinsky's Settings of English', *The Score*, no.20 (1957), 14

R. Sessions: 'Thoughts on Stravinsky', *The Score*, no.20 (1957), 32

L. Kerstein: 'Purity through the Will', *The Nation*, no.191 (8 Oct 1960), 233

B. Schwarz: 'Stravinsky in Soviet Russian Criticism', *MQ*, xlviii (1962), 340

T. W. Adorno: 'Stravinsky: ein dialektisches Bild', *Quasi una fantasia* (Frankfurt, 1963)

V. Bogdanov-Berezovsky: 'Igor' Stravinskii i evo "Khronika"', introduction to Russ. edn. of I. Stravinsky: *Chroniques de ma vie* (Leningrad, 1963)

M. Boykan: 'Neoclassicism and Late Stravinsky', *PNM*, i/2 (1963), 155

H. Curjel: 'Strawinsky und die Maler', *Strawinsky: ein Sendereihe des Westdeutschen Rundfunks zum 80. Geburtstag* (Cologne, 1963); repr. in *Melos*, xxxiv (1967), 203

V. Duke [V. Dukelsky]: 'The Deification of Stravinsky', *Listen Here!* (New York, 1963), 149–89

P. H. Lang, ed.: *Stravinsky: a New Appraisal of his Work* (New York, 1963), 9 [Introduction]

R. Shattuck: 'Making Time: a Study of Stravinsky, Proust and Sartre', *Kenyon Review*, xxv (1963), 248

E. Ansermet: 'The Crisis of Contemporary Music: II. The Stravinsky Case', *Recorded Sound*, no.14 (1964), 197

G. Shneyerson: *O muzïke zhivoy i myortvoy* [On music alive and dead] (Moscow, 2/1964), 211–46

Bibliography

U. Dibelius: 'Strawinskys musikalische Wirklichkeit', *Melos*, xxxiv (1967), 189

'Personal Viewpoints: Notes by Five Composers', *Tempo*, no.81 (1967), 19

E. Roth: 'Igor Strawinsky: sein Weg und sein Werk für die moderne Musik', *Universitas*, xxii (1967), 839

P. Souvtchinsky and J. Warrack: 'Stravinsky as a Russian', *Tempo*, no.81 (1967), 5

L. Davies: 'Stravinsky as Littérateur', *ML*, xlix (1968), 135

H. A. Leibowitz, ed: *Musical Impressions: Selections from Paul Rosenfeld's Criticism* (London, 1969), 126–78

A. W. Whittall: 'Stravinsky and Music Drama', *ML*, i (1969), 63

A. Berger: 'Neoclassicism Reexamined', *PNM*, ix/2–x/1 (1971), 79

S. D. Burton and others: 'Die junge Generation äussert sich über Strawinsky', *Melos*, xxxviii (1971), 348

H. Fiechtner: 'Keine Kunst ohne Kanon. "Zar Igor". Über das Werk Strawinskys', *Marginalien zur poetischen Welt: Festschrift für Robert Mühler zum 60. Gerburtstag* (Berlin, 1971), 489

D. Harris: 'Stravinsky and Schoenberg: a Retrospective View', *PNM*, ix/2–x/1 (1971), 108

A. Huber: 'Adornos Polemik gegen Strawinsky', *Melos*, xxxviii (1971), 356

J. Noble: 'Igor Stravinsky, 1882–1971', *MT*, cxii (1971), 534

H. Pousseur: 'Stravinsky selon Webern selon Stravinsky', *Musique en jeu*, no.4 (1971), 21; no.5 (1971), 107; Eng. trans. in *PNM*, x/2 (1972), 13–51; xi/1 (1972), 112–45

H. H. Stuckenschmidt: 'Igor Strawinsky und sein Werk für die moderne Musik', *Universitas*, xxvi (1971), 471

D. Bancroft: 'Stravinsky and the "NRF"', *ML*, liii (1972), 274; lv (1974), 261

M. Druskin: 'Zametki o Stravinskom' [Notes on Stravinsky], *SovM* (1972), no.8, p.83; Ger. trans., *Kunst und Literatur*, xxi (1973), 830

N. Nabokov: 'Igor Stravinsky, a Partisan Chronicle', *Coloquio artes*, x (1972), 15

H. H. Stuckenschmidt: 'Igor Strawinsky und seine Wandlungen', *Universitas*, xxvii (1972), 613

R. Middleton: 'Stravinsky's Development: a Jungian Approach', *ML*, liv (1973), 289

A. Shnitke: 'Paradoksal'nost' kak cherta muzïkal'noy logiki Stravinskovo' [Paradoxicality as a feature of Stravinsky's musical logic], *Stat'i i materialï*, compiled L. S. D'yachkova (Moscow, 1973), 383–434

S. Slonimsky: 'Ob I. F. Stravinskom' [On I. F. Stravinsky], *Stat'i i materialï*, compiled L. S. D'yachkova (Moscow, 1973), 105

N. G. Shakhnazarova: *Problemï muzykal'noy estetiki v teoreticheskikh trudakh Stravinskovo, Schenberga, Khindemita* [Problems of musical aesthetics in the theoretical works of Stravinsky, Schoenberg and Hindemith] (Moscow, 1975)

R. Craft: 'Stravinsky, Roland-Manuel, and the "Poetics of Music" ', *The Stravinsky Festival* (London, 1979) [special suppl. to programme]

A. Fiorenza: 'Stravinsky tra Adorno e Varese', *NRMI*, xiii (1979), 759

J. Viret: 'Stravinsky, 'mauvais génie' du ballet contemporain?', *Revue Musicale de Suisse Romande*, xxxiv (1981), 190

M. Druskin: 'Werk und Persoenlichkeit Igor Strawinskys; zum 100. Geburtstag des Komponisten am 17. Juni 1982', *Musik und Gesellschaft*, xxxii (1982), 321

M. Hansen: 'Apollons Befehl; Anmerkungen zum Kunstbegriff Igor Strawinskys', *Musik und Gesellschaft*, xxxii (1982), 335

D. Harris: 'Balanchine on Stravinsky', *Keynote*, vi/4 (1982), 15

T. Hirsbrunner: 'Strawinsky und die Antike', *ÖMz*, xxxvii (1982), 320

A. Kuznetsov: 'V zerkale russkoy kritiki' [In the mirror of Russian criticism], *SovM* (1982), no.6, p.69

H. Lindlar: 'Noten-Nachlese zu Strawinsky', *ÖMz*, xxxvii (1982), 356

E. Schönberger and L. Andriessen: 'The Apollonian Clockwork: Extracts from a Book', *Tempo*, no.141 (1982), 3

J. Pasler: 'Stravinsky and his Craft: Trends in Stravinsky Criticisms and Research', *MT*, cxxiv (1983), 605

ANALYTICAL STUDIES

B. V. Asaf'yev: 'Prozess der Formbildung bie Stravinsky', *Der Auftakt*, ix (1929), 101

O. Messiaen: 'Le rythme chez Igor Stravinsky', *ReM*, no.191 (1939), 91

D. Drew: 'Stravinsky's Revisions', *The Score*, no.20 (1957), 47

R. Gerhard: 'Twelve-note Technique in Stravinsky', *The Score*, no.20 (1957), 38

H. Keller: 'Rhythm: Gershwin and Stravinsky', *The Score*, no.20 (1957), 19

R. Travis: 'Towards a New Concept of Tonality?', *JMT*, iii (1959), 257–94

E. T. Cone: 'The Uses of Convention: Stravinsky and his Models', *MQ*, xlviii (1962), 287

D. C. Johns: 'An Early Serial Idea of Stravinsky', *MR*, xxiii (1962), 305

D. Mitchell: 'Stravinsky and Neoclassicism', *Tempo*, no.61–2 (1962), 9

Bibliography

R. U. Nelson: 'Stravinsky's Concept of Variations', *MQ*, xlviii (1962), 327

Yu. Kholopov: 'Formoobrazuyushchaya rol' sovremennoy garmonii' [The form-determining role of contemporary harmony], *SovM* (1965), no.11, p.47

G. W. Hopkins: 'Stravinsky's Chords', *Tempo* (1966), no.76, p.6; no.77, p.2

V. N. Kholopova: 'O ritmicheskoy tekhnike i dinamicheskikh svoystvakh ritma Stravinskovo' [On Stravinsky's rhythmic technique and the dynamic properties of his rhythm], *Muzïka i sovremennost'*, iv (1966)

W. Mellers: 'Stravinsky and Jazz', *Tempo*, no.81 (1967), 29

V. Smirnov: 'O predposïlkakh evolyutsii Stravinskovo k neoklassitsismu' [On the premises for Stravinsky's evolution towards neoclassicism], *Voprosï teorii i estetiki muzïki*, v (1967), 142

A. Berger: 'Problems of Pitch Organization in Stravinsky', *Perspectives on Schoenberg and Stravinsky*, ed. B. Boretz and E. T. Cone (Princeton, NJ, 1968, 2/1972), 123

G. B. Biggs: *The Return Effect in Works for Orchestra by Stravinsky, Bartók, and Schoenberg, as Determined by Factors other than Theme and Key* (diss., U. of Indiana, 1968)

E. T. Cone: 'Stravinsky: the Progress of a Method', *Perspectives on Schoenberg and Stravinsky*, ed. B. Boretz and E. T. Cone (Princeton, NJ, 1968, 2/1972), 156

L. Ivanova: 'Stravinsky i traditsiya' [Stravinsky and tradition], *SovM* (1968), no.5, p.85

G. Grigor'yeva: 'Russkiy fol'klor v sochineniyakh Stravinskovo' [Russian folklore in Stravinsky's works], *Muzïka i sovremennost'*, vi (1969)

L. S. D'yachkova: 'Politonal'nost' v tvorchestve Stravinskovo' [Polytonality in the work of Stravinsky], *Voprosï teorii muzïki*, ii, ed. Yu. N. Tyulin (Moscow, 1970)

U. Kraemer: 'Das Zitat bei Igor Strawinsky', *NZM*, Jg.131 (1970), 135

G. J. Mokreyeva: [On the melodies of Stravinsky], *Voprosï teorii muzïki*, ii, ed. Yu. N. Tyulin (Moscow, 1970)

A. Berger: 'Neoclassicism Reexamined', *PNM*, ix/2–x/1 (1971), 79

R. Moevs: 'Mannerism and Stylistic Consistency in Stravinsky', *PNM*, ix/2–x/1 (1971), 92

K. Oppens: 'Die Dialektik in der Musik Strawinskys', *Merkur*, xxv (1971), 930

J. Hunkemöller: 'Igor Strawinskys Jazz-Porträt', *AMw*, xxix (1972), 45

R. Maconie: 'Stravinsky's Final Cadence', *Tempo*, no.103 (1972), 18

C. Mason: 'Stravinsky's Contributions to Chamber Music', *Tempo*, no.43 (1957), 6

H. Lindlar: *Igor Stravinskys sakraler Gesang* (Regensburg, 1957)

L. Erhardt: *Balety Igor Stravinskego* (Kraków, 1962)

J. Ogdon: 'Stravinsky and the Piano', *Tempo*, no.81 (1967), 36

S. Walsh: 'Stravinsky's Choral Music', *Tempo*, no.81 (1967), 41

M. Zijlstra: 'De liederen van Strawinsky', *Mens en melodie*, xxv (1970), 11

V. Krasovskaya: *Russkiy baletniy teatr nachala XX veka* [The Russian ballet theatre at the beginning of the 20th century], i (Leningrad, 1971), 352, 367, 420

S. H. Forster: *The Solo Piano Works of Igor Stravinsky* (diss., U. of Indiana, 1972)

L. Klein: 'Stravinsky and Opera: Parable as Ethic', *Les cahiers canadiens de musique*, no.4 (1972), 65

C. M. Joseph: *A Study of Igor Stravinsky's Piano Compositions* (diss., U. of Cincinnati, 1974)

H. Kirchmeyer: *Stravinskys russische Ballette* (Stuttgart, 1974)

W. E. McCandless: *Cantus firmus Techniques in Selected Instrumental Compositions, 1910–1960* (diss., U. of Indiana, 1974)

V. Thomson: 'Stravinsky's Operas', *Musical Newsletter*, iv/4 (1974), 3

A. Vul'fson: *Printsipï simfonicheskovo razvitiya i formoobrazovaniya v baletakh I. F. Stravinskovo* [Principles of symphonic development and form building in I. F. Stravinsky's ballets] (diss., Leningrad Conservatory, 1974)

M. Thal: *The Piano Music of Igor Stravinsky* (diss., U. of Washington, 1978)

R. Zinar: 'Stravinsky and his Latin Texts', *College Music Symposium*, xviii (1978), 186

T. Hirsbrunner: 'Zu Strawinskys Kammermusik', *Musica*, xxxvi (1982), 383

WORKS: EARLY PERIOD
(*collective*)

H. Lindlar; 'Debussysmen beim frühen Strawinsky', *Bericht über den Internationalen Musikwissenschaftlichen Kongress, Kassel 1962* (Kassel, 1963), 252

G. Mokreyeva: 'Ob evolyutsii garmonii rannevo Stravinskovo' [On the evolution of harmony in early Stravinsky], *Teoricheskiye problemï muzïki XX veka*, i, ed. Yu. N. Tyulin (Moscow, 1967)

A. Shnitke: 'Osobennosti orkhestrovovo golosovedeniya rannikh proizvedeniy Stravinskovo' [Characteristics of the orchestral part-writing in Stravinsky's early works], *Muzĭka i sovremennost'*, v (1967), 209

I. Vershinina: *Ranniye baletĭ Stravinskovo* [Stravinsky's early ballets] (Moscow, 1967)

V. Smirnov: 'U istokov kompozitorskovo puti Igor Stravinskovo' [On the wellsprings of Stravinsky's path as a composer], *Voprosĭ teorii i estetiki muziki*, viii (1968)

I. Kovshar: *Igor' Stravinsky: ranneye tvorchestvo* [Igor Stravinsky: early works] (Baku, 1969)

H. Lindlar: 'Die frühen Lieder von Strawinsky', *Musica*, xxiii (1969), 116

V. V. Smirnov: *Tvorcheskoye formirovaniye I(gorya) F(yodorovicha) Stravinskovo* [Stravinsky's formative period] (Leningrad, 1970)

G. Cowart: *The Early Works of Igor Stravinsky* (diss., U. of Indiana, 1972)

C. Huston: *A Study of Igor Stravinsky's Early Works* (diss., Florida State U., 1972)

G. Pestelli: *Il giovane Stravinsky (1906–13)* (Turin, 1973)

V. Smirnov: 'Tvorcheskaya vesna Igoriya Stravinskovo' [Igor Stravinsky's creative springtime], *Rasskazĭ o muzĭke i muzĭkantakh*, ed. M. G. Aranovich (Leningrad, 1973), 55

R. Taruskin: 'From *Firebird* to *The Rite*: Folk Elements in Stravinsky's Scores', *Ballet Review*, x (1982), 72

S. Karlinsky: 'Stravinsky and Pre-literate Russian Theater', *19th-century Music*, vi (1983), 232

R. Taruskin: 'From Subject to Style: Stravinsky and the Painters', *Stravinsky: Centennial Essays*, ed. J. Pasler (Berkeley, Calif., in preparation)

(Scherzo)

C. M. Joseph: 'Stravinsky's Piano Scherzo (1902) in Perspective: a New Starting Point', *MQ*, lxvii (1981), 82

(Symphony in E♭ major)

T. L. Greenwalt: *A Study of the Symphony in Russia from Glinka to the Early 20th Century* (diss., Eastman School of Music, U. of Rochester, 1972)

(Feu d'artifice)

M. Just: 'Tonordnung und Thematik in Strawinskys "Feu d'artifice", op.4', *AMw*, xl (1983), 61

(Four Studies)

J. Orvis: *Technical and Stylistic Features of the Piano Etudes of Stravinsky, Bartók, and Prokofiev* (diss., U. of Indiana, 1974)

wiosny' [The harmonic thinking in Stravinsky's *Rite of Spring*], *Res facta*, v (1971), 104

A. Schaeffner: 'Au fil des esquisses du "Sacre"', *RdM*, lvii (1971), 179

J. Siddons: 'Rhythmic Structures in *Le sacre du printemps* (danse sacrale)', *Musical Analysis*, i (1972), 6

J. W. McCalla: *Structural and Harmonic Innovations in the Music of Schoenberg, Stravinsky and Ives prior to 1915* (diss., New England Conservatory of Music, 1973)

A. Vul'fson: [The symphonic idiom in Stravinsky's *Rite of Spring*], *Stranitsï istorii russkoy muzïki*, ed. E. Orlova and E. Ruts'evskaya (Leningrad, 1973)

B. Yarustovsky: 'I. Stravinsky: eskiznaya tetrad (1911–1913); nekotorïye nabliudeniya i razmyshleniya' [I. Stravinsky: a sketchbook (1911–1913): some observations and reflections], *Stat'i i materialï*, compiled L. S. D'yachkova (Moscow, 1973), 162–206

A. Jarzębska: 'Pierre Bouleza koncepja analityczna *Swioęta wiosny* Strawińskiego' [Pierre Boulez's analytic conception of Stravinsky's *Rite of Spring*], *Muzyka*, xx/2 (1975), 47

R. Craft: ' "Le sacre du printemps": the Revisions', *Tempo*, no.122 (1977), 2

——: 'Craft on Forte', *MQ*, lxiv (1978), 524 [review of Forte, 1978]

A. Forte: *The Harmonic Organization of "The Rite of Spring"* (New Haven, 1978)

T. Bullard: 'The Riot at the *Rite*: not so Surprising After All', *Essays on Music for Charles Warren Fox*, ed. J. Graue (Rochester, NY, 1979), 206

R. Craft: '*Le Sacre du printemps:* a Chronology of the Revisions', *The Stravinsky Festival* (London, 1979) [special suppl. to programme]

L. Morton: 'Footnotes to Stravinsky Studies: *Le Sacre du printemps*', *Tempo*, no.128 (1979), 9

R. Taruskin: review of A. Forte: *The Harmonic Organization of The Rite of Spring*, *CMc*, no.28 (1979), 114

J. J. Eigeldinger: 'Une lettre inédite d'Ansermet à Stravinsky à propos du *Sacre du printemps*', *Revue Musicale de Suisse Romande*, xxxiii/5 (1980), 10

M. Hodson: 'Searching for Nijinsky's *Sacre*', *Dance Magazine*, xiv/6 (1980), 64, 71

F. Lesure, ed.: *Igor Stravinsky: Le sacre du printemps, dossier de presse* (Geneva, 1980)

R. Moevs: review of A. Forte: *The Harmonic Organization of The Rite of Spring*, *JMT*, xxiv (1980), 97

R. Taruskin: 'Russian Folk Melodies in *The Rite of Spring*', *JAMS*, xxxiii (1980), 501–43

Bibliography

S. Walsh: 'Forte in his own *Rite*', *Soundings*, no.9 (1980–81), 81 [review of Forte, 1978]

A. Whittall: 'Music Analysis as Human Science? *Le sacre du printemps* in Theory and Practice', *Music Analysis*, i (1982), 33

R. Taruskin: 'The *Rite* Revisited: the Idea and the Sources of its Scenario', *Music and Civilization: Essays in Honor of Paul Henry Lang*, ed. M. R. Maniates and E. Strainchamps (New York, in preparation)

(*Le rossignol*)

M. Calvocoressi: 'Igor Stravinsky's Opera "The Nightingale"', *MT*, lv (1914), 372

N. Kinkul'kina: 'Pis'ma I. F. Stravinskovo i F. I. Shalyapina k A. A. Saninu' [Stravinsky's and Shalyapin's letters to Sanin], *SovM* (1978), no.6, p.92

(*Three Pieces*)

W. Kolneder: 'Strawinskys *Drei Stücke für Streichquartett*', *ÖMz*, xxvi (1971), 631

R. Stephan: 'Aus Igor Strawinskys Spielzeugschachtel', *Erich Doflein Festschrift* (Mainz, 1972), 27

C. J. Smith: 'Comment on Gryč', *In Theory Only*, ii/3–4 (1976), 7

(*Pribaoutki*)

B. Bartók: 'The Influence of Folk Music in the Art Music of Today', *Béla Bartók Essays*, ed. B. Suchoff (New York, 1976), 316

(*Histoire du soldat*)

C. F. Ramuz: 'La naissance de *L'histoire du soldat*', *Lettres, 1900–18* (Lausanne, 1956)

S. Bradshaw: '*Soldier's Tale* Suite', *Tempo*, no.97 (1971), 15

J. Jacquot, ed.: *Les voies de la création théâtrale, 6: Théâtre et musique: histoire du soldat par Ramuz et Stravinsky* (Paris, 1978)

M. Trapp: *Studien zu Strawinskys 'Geschichte vom Soldaten' (1918): zur Idee und Wirkung des Musiktheaters der 1920er Jahre* (Regensburg, 1978)

R. Craft: '*Histoire du Soldat*: (the Musical Revisions, the Sketches, the Evolution of the Libretto)', *MQ*, lxvi (1980), 321

T. Moshonkina: 'Grotesk v "Istorii soldata" Stravinskovo' [The grotesque in Stravinsky's 'Histoire du Soldat'], *O Muzïke*, ed. E. P. Fedosova (Moscow, 1980), 40

C. Marti: 'Zur Kompositionstechnik von Igor Strawinsky; das *Petit Concert* aus der *Histoire du Soldat*', *AMw*, xxxviii (1981), 93

A. Traub: *Igor Strawinsky, 'L'histoire du soldat'* (Munich, 1981)

(*Piano-rag-music*)

C. M. Joseph: 'Structural Coherence in Stravinsky's *Piano-rag-music*', *Music Theory Spectrum*, iv (1982), 76

217

G. Vinay: 'Da "Oedipus" a "Oedipus rex" e ritorno: un itinerario metrico', *RIM*, xvii (1982), 333

(*Apollon musagète*)

D. Gutknecht: 'Strawinskys zwei Fassungen des "Apollon musagète" ', *Musicae scientiae collectanea: Festschrift Karl Gustav Fellerer* (Cologne, 1973), 199

(*Le baiser de la fée*)

L. Morton: 'Stravinsky and Tchaikovsky: "Le baiser de la fée" ', *MQ*, xlviii (1962), 313; repr. in *Stravinsky: a New Appraisal of his Work*, ed. P. H. Lang (New York, 1963), 47

M. Mikhaylov: 'Esteticheskiy fenomen *Potseluya feï* [The aesthetic phenomenon of *Le baiser de la fée*], *SovM* (1982), no.8, p.95

(*Symphony of Psalms*)

W. Piston: 'Stravinsky as Psalmist', *MM*, xiii (1931), 42

D. R. Chittum: 'Compositional Similarities in Beethoven and Stravinsky', *MR*, xxx (1969), 285

W. Mellers: 'Symphony of Psalms', *Tempo*, no.97 (1971), 19

G. Alfeyevskaya: 'O *Simfonii psalmov* I. Stravinskovo', *Stat'i i materialï*, compiled L. S. D'yachkova (Moscow, 1973), 250

R. K. Debruyn: *Contrapuntal Structure in Contemporary Tonal Music: a Preliminary Study of Tonality in the Twentieth Century* (diss., U. of Illinois, 1975)

(*Violin Concerto*)

E. Pałłasz: Elementy barokowego stylu w Koncercie skrzypcowym D-dur I. Strawińskiego' [Elements of the Baroque style in Stravinsky's Violin Concerto in D], *Polsko-rosyjskie miscellanea muzyczne* (Kraków, 1967), 297

J. Little: *Architectonic Levels of Rhythmic Organization in Selected Twentieth-century Music* (diss., U. of Indiana, 1971)

(*Symphony in C*)

S. Babitz: 'Stravinsky's Symphonie in C (1940)', *MQ*, xxvii (1941), 20

B. M. Williams: 'Time and the Structure of Stravinsky's Symphony in C', *MQ*, lix (1973), 355

P. G. Johnson: *The First Movement of Stravinsky's 'Symphony in C': its Syntactical Bases and their Implications* (diss., Princeton U., 1981)

(*Four Norwegian moods*)

U. Kraemer: '*Four Norwegian moods* von Igor Strawinsky', *Melos*, xxxix (1972), 80

K. Velten: 'Igor Strawinskys *Four Norwegian Moods* im Unterricht', *Musik und Bildung*, xi (1979), 246

Bibliography

(Sonata for two pianos)

D. Johns: 'An Early Serial Idea of Stravinsky', *MR*, xxiii (1962), 305

C. Burkhart: 'Stravinsky's Revolving Canon', *MR*, xxix (1968), 161

(Orpheus)

I. Dahl: 'The New Orpheus', *Stravinsky in the Theatre*, ed. M. Lederman (London and New York, 1949/*R*1975), 70

(Mass)

R. Craft: 'Stravinsky's Mass: a Notebook', *Igor Stravinsky*, ed. E. Corle (New York, 1949), 201

D. V. Moses: *A Conductor's Analysis of the Mass (1948) by Stravinsky* (diss., U. of Indiana, 1968)

E. Malïsheva: 'O gruzinskikh istokakh Messï' [On the Georgian sources of the Mass], *SovM* (1982), no.7, p.92

B. L. Vantine: *Four Twentieth-century Masses: an Analytical Comparison of Style and Compositional Technique* (diss., U. of Illinois, 1982)

(The Rake's Progress)

C. Mason: 'Stravinsky's Opera', *ML*, xxxiii (1952), 1

I. Stravinsky and R. Craft: 'The Rake's Progress', SL-125 [disc notes]

J. Kerman: 'Opera à la mode', *Hudson Review*, vi (1953–4), 560

R. Craft: 'Reflections on *The Rake's Progress*', *The Score*, no.9 (1954), 24

D. Cooke: '*The Rake* and the 18th century', *MT*, ciii (1962), 20

A. A. Abert: 'Strawinsky's *The Rake's Progress*: strukturell betrachtet', *Musica*, xxv (1971), 243

G. Josipovici: '*The Rake's Progress*: Some Thoughts on the Libretto', *Tempo*, no.113 (1975), 2

G. Ordzhonikidze: 'Nravstvennye uroki nasmeshlivoy pritchi' [Moral lessons of a mocking parable], *SovM* (1979), no.1, p.56

P. Griffiths: *Igor Stravinsky: The Rake's Progress* (Cambridge, 1982)

J. Jacquot: 'The Rake's Progress et la carrière de Stravinsky', *RdM*, lxviii (1982), 110

WORKS: LATE SERIAL PERIOD
(collective)

D. Ward-Steinman: *Serial Technique in the Recent Music of Igor Stravinsky* (diss., U. of Illinois, 1961)

M. Babbitt: 'Remarks on the Recent Stravinsky', *Perspectives on Schoenberg and Stravinsky*, ed. B. Boretz and E. T. Cone (Princeton, NJ, 1968, 2/1972), 165

——: 'Contemporary Music Composition and Music Theory as Contemporary Intellectual History', *Perspectives in Musicology* (New York, 1972), 151

(*Requiem Canticles*)

A. Payne: 'Requiem Canticles', *Tempo*, no.81 (1967), 10

E. Salzman: 'Current Chronicle: United States, Princeton', *MQ*, liii (1967), 589

C. Spies: 'Some Notes on Stravinsky's Requiem Settings', *Perspectives on Schoenberg and Stravinsky*, ed. B. Boretz and E. T. Cone (Princeton, NJ, 1968, 2/1972), 223

P. Souvtchinsky: 'Thoughts on Stravinsky's *Requiem Canticles*', *Tempo*, no.86 (1968), 6

M. Zijlstra: 'Strawinsky's *Requiem Canticles*: een nieuw geestelijk koorwerk', *Mens en melodie*, xxiii (1968), 76

ARRANGEMENTS

R. Craft: 'A Note on Gesualdo's 'Sacrae cantiones' and on Gesualdo and Stravinsky', *Tempo*, no.45 (1957), 5

C. Mason: 'Stravinsky and Gesualdo', *Tempo*, no.55–6 (1960), 39

P. Nielsen: 'Om Johann Sebastian Strawinsky', *Dansk musiktidsskrift*, xliii/5 (1967), 100

R. Threlfall: 'The Stravinsky version of *Khovanshchina* in collaboration with Ravel', *SMA*, no.15 (1981), 106

T. van Huijstee: 'Stravinsky's *Vom Himmel Hoch*', *Mens en melodie*, xxxvi (1981), 172

H. Lindlar: 'Zu Strawinskys geistlichem Vermaechtnis; um *zwei Gesaenge aus Wolfs Spanischem Liederbuch*', *ÖMz*, xxxvii (1982), 318

N. Jers: 'Bearbeitungen in Strawinskys Spätwerke', *Musica*, xxxvii (1983), 239

MISCELLANEOUS

G. Stempowski: 'Auf Igor Strawinskys Spuren in Wolhynien', *Du*, xii (Zurich, 1950), 33

P. Meylan: *Une amitié célèbre: C. F. Ramuz/Igor Stravinsky* (Lausanne, 1961)

E. W. White: 'Stravinsky and Debussy', *Tempo*, no.61–2 (1962), 2

J. Noble: 'Debussy and Stravinsky', *MT*, cviii (1967), 22

J. Spiegelman: 'Stravinsky and the New Russians', *PNM*, ix/2–x/1 (1971), 126

H. Curjel: 'Strawinsky in Berlin,' *Melos*, xxxix (1972), 154

R. Craft: 'Stravinsky: Relevance and Problems of Biography', *Prejudices and Disguises* (New York, 1974), 270

W. S. Russell: *Stravinsky and Eliot: Personality, Poetics and Cultural Politics* (diss., Emory U., 1976)

M. M. Boaz: *T. S. Eliot and Music: a Study of the Development of Musical Structures in Selected Poems by T. S. Eliot and Music by*

Bibliography

Erik Satie, Igor Stravinsky and Béla Bartók (diss., U. of Illinois, 1977)

J. Pasler: 'Stravinsky and the Apaches', *MT*, cxxiii (1982), 403

J. Peyser: 'Stravinsky–Craft, Inc.', *American Scholar*, lii (1983), 513

PAUL HINDEMITH

Ian Kemp

CHAPTER ONE

Life

Paul Hindemith was born in Hanau, near Frankfurt, on 16 November 1895. His father, Robert Rudolf Emil Hindemith, came from Lower Silesia; his mother, Marie Sophie Warnecke, from Hesse. In 1905 Robert Rudolf, who was a house-painter, settled in Frankfurt, where he set up a small business in a working-class district. An enthusiastic music lover who played the zither, he had, in his teens, run away from home because his father would not allow him to become a musician. Robert Rudolf was determined therefore to make musicians of his own children, not least because he thought this would improve their social status. From an early age he subjected them to a strict discipline of practice and training. Hindemith was the eldest of three children (the others were Toni and Rudolf, a sister and a brother). During his childhood the family lived in real poverty and relationships were strained, the atmosphere joyless. Hindemith became extremely reticent about his working-class origins, which he seems to have regarded as both an embarrassment and an irrelevance. When his astonishing gifts attracted the attention of several well-to-do Frankfurt families, he ensured that he was accepted for what he could do, not for what he represented. His subsequent insistence on skill, craftsmanship, on writing for amateurs, shows how anxious he was to keep sentiment at a distance.

Hindemith began violin lessons at the age of six, and in 1907 he became a pupil of Anna Hegner, who taught at the Hoch Conservatory, Frankfurt. Before leaving Frankfurt in 1908 Hegner introduced Hindemith to Adolf Rebner, the doyen of violin teachers at the conservatory, who immediately accepted the 12-year-old boy as a private pupil and in the following year arranged that Hindemith be awarded a free place at the conservatory. Hindemith was a student there until 1917. During the first three years he concentrated on the violin; then in 1912 he began studies in composition, initially with Arnold Mendelssohn and subsequently with Bernard Sekles. Hindemith's exceptional instrumental gifts (he was to become an accomplished performer on several instruments, notably the clarinet and the piano, as well as the violin and the viola) developed so rapidly that in 1915, when he was 19, he not only took over the second violin in Rebner's string quartet but was also appointed a first violinist (in June) and then leader (in September) of the Frankfurt Opera Orchestra. In the same year Hindemith's father, who had volunteered for military service, was killed in Flanders. Hindemith had already contributed to household finances, by playing in café orchestras and similar groups, but his income now became a necessity for supporting his family. In 1917 he was called up himself. He spent most of his service as a member of a regimental band about 3 km from the front line, and was fortunate in having a sympathetic commanding officer who asked him to form a string quartet to give private concerts (see fig.21). After the war Hindemith returned to the Rebner Quartet (from 1919 he transferred to the viola) and to the Frankfurt Opera Orchestra. In 1924 he mar-

230

ried Gertrud, daughter of the conductor of the Opera, Ludwig Rottenberg.

Hindemith had begun writing music long before becoming a student. At the conservatory some of his works were performed at concerts given by members of Sekles's composition classes. By the time of the first public concert of his music he had already developed a style of some individuality. This concert was on 2 June 1919 at the newly founded Verein für Theater- und Musikkultur in Frankfurt, when his Piano Quintet op.7, two sonatas from op.11 and the String Quartet op.10 were performed. Hindemith now attracted attention; his music was published. In June 1921 the premières at Stuttgart of *Mörder, Hoffnung der Frauen* and *Das Nusch-Nuschi*, the first two of his three early one-act operas, created minor scandals (because of their provocative attitude towards sexuality) and in August of that year his String Quartet no.2 op.16 was performed at the first Donaueschingen Festival. With this latter performance his reputation as the leading young composer in Germany was securely established; it was consolidated by the vociferous reception given at the 1922 festival to the very different *Kammermusik no.1*.

Various features of the *Kammermusik no.1* suggest that Hindemith's motives at that time were almost dadaistic, but in fact he was more concerned with questioning received ideas about the conventional concert. Shortly before the first performance of the *Kammermusik* he had formed a concert 'community' in Frankfurt that would be uncommercial, would invite no critics, would not announce the names of the performers, and would perform unknown music, new and old. This project met with little success, but Hindemith

21. *Paul Hindemith (extreme left) with his soldiers' string quartet during World War I*

had given notice of his intention to lead concert-giving away from its familiar ritual and direct it towards more pointed ends. He was able to put some of these ideas into practice when, in 1923, he was invited to join the administrative committee of the Donaueschingen Festival. Along with Heinrich Burkard (the creator of the festival) and Joseph Haas, he guided its fortunes, and it soon developed an importance out of all proportion to its size, thanks largely to Hindemith's energy and imagination. Between 1923 and 1924 he included none of his own music but exerted an increasing control. In 1924 Schoenberg's op.24, Webern's opp.9 and 14, and pieces for quarter-tone piano were performed, and in subsequent years Hindemith decided to concentrate on particular themes rather than dilute the programmes with inferior submitted works. Thus the festival included specially commissioned music for unaccompanied chorus, military band, films and mechanical instruments, chamber operas, and music for amateurs. In 1927 the festival was transferred to Baden-Baden, where it could present more ambitious projects such as the chamber operas, and in 1930 to Berlin, where it appeared for the last time before its revival after World War II.

In 1927 Hindemith accepted an appointment as professor of composition at the Hochschule für Musik in Berlin. At the time this was a surprising decision, since he had no teaching experience and appeared to be committed to practical music. But with relatively small classes and talented pupils, all of whom had already acquired some musical qualification, he quickly developed a lasting interest in teaching and devoted great care to it. As well as teaching at the Hochschule he

233

also took an evening class at the Volksmusikschule Neu-Kölln in Berlin. His experience with music students and amateur players led both to the writing of the theoretical work *Unterweisung im Tonsatz* and the composition of music for amateurs.

During this period Hindemith also led an active life as a performer. The Amar–Hindemith Quartet (Licco Amar, Walter Kaspar, Hindemith and, initially, his brother Rudolf, and then Maurits Frank) was formed in 1921 simply in order that the Quartet no.2 could be heard (the Havemann Quartet had been engaged to perform the work but refused to do so because they found it too difficult), but it was rapidly acknowledged as one of the foremost quartets in Europe, especially admired for its performances of contemporary music. The success of the quartet obliged Hindemith to give up his position with the Frankfurt Opera Orchestra in 1923. The quartet ceased playing in 1933; Hindemith himself had left in 1929, when he had found that he could not devote sufficient time to it. In that year, he formed a string trio with Josef Wolfsthal and Emanuel Feuermann, who also taught at the Hochschule. In 1931 Wolfsthal died; his place was taken by Szymon Goldberg, and the trio continued to give concerts until 1934, when it was forced to disband because its members could not meet regularly. During these years Hindemith frequently performed as a soloist, both on the viola (notably in the first performance of Walton's Viola Concerto at a Promenade Concert on 3 October 1929) and the viola d'amore, and he was widely regarded as the most accomplished and versatile performer–composer of his time. This versatility and, especially, his compositional fertility became almost legendary. He seemed to write a composition as easily as a letter; his *Kleine Kammermusik* was written in five days, the finale of the Solo Violin Sonata op.31 no.1 on the train between Bremen and Frankfurt, the *Trauermusik* the day after the death of George V (and it was performed the day after that).

When the Nazis came to power they did not immediately seek to discredit Hindemith, even though sections of the musical press had been complaining since 1930 that he was betraying his mission as a German composer. Hindemith himself adopted a pragmatic approach, bending a little towards the demands of a regime he thought would be short-lived. His roots were in Germany, he had made his name there, and he did not want to leave it, whoever was in charge. He seemed oddly unmoved by the dismissal of Jewish musicians from the Hochschule; yet he continued playing with Jews, collaborated with the Jewish Kulturbund in a performance of his children's opera *Wir bauen eine Stadt* and made no secret of his anti-Nazi views to his composition class. After the event, it can be said that he was naive rather than opportunist. In the summer of 1934 a campaign was launched against him, based on his membership of an 'international' group of atonal composers, the supposed immorality of his one-act operas, his 'parody' in the finale of the *Kammermusik no.5* (1927) of a Bavarian military march heard at Nazi rallies, and in particular his association with Jews. It may have been provoked by some uncomplimentary remarks he made about Hitler, when on a visit to Switzerland. In November 1934 the Kulturgemeinde (a semi-official organization that had assumed responsibility for the spiritual welfare of the Nazi party) announced a boycott on performances of

Hindemith's music. This greatly angered Furtwängler (the conductor in March 1934 of the strikingly successful first performance of the symphony *Mathis der Maler*), who wrote an article for the *Deutsche allgemeine Zeitung* of 25 November stoutly defending the composer. At the Berlin Staatsoper that evening Furtwängler was not allowed to begin the performance until he had received an ovation lasting 20 minutes. It was clear that Hindemith and everything he stood for posed a serious threat to the Nazis. Furtwängler was relieved of his conducting and administrative posts and in December Goebbels made a personal attack on Hindemith at a Nazi rally. In January 1935 Hindemith was given a six-month 'leave of absence' from the Hochschule. But he was not expelled from the country. He evidently felt that it was still worth trying to reach some sort of accommodation and approved a letter from his publishers representing him as a model composer for the new state. In mid-1936, he did receive encouragement, in the form of a commission from the Luftwaffe, which he accepted (though never fulfilled), and a concert including his recently written Violin Sonata in E was permitted. The sonata however was received with a warmth that could not have been directed wholly at the music and with that hopes of rehabilitation came to an end. Some performances of his music did occur – the Kulturgemeinde boycott was not endorsed by the Reichsmusikkammer (the music division of the Nazi Ministry of Culture) until 1937, there being some rivalry between the two organizations – and he was allowed to return to teaching at the Hochschule, to undertake concert tours abroad (notably to the USA), to enter into an agreement with the Turkish government to build up an organized musical life in that country, and to have his music published. In 1937, however, he gave up his teaching post and in the following year left Germany, settling in Bluche ob Sierre in Switzerland. In a letter of 20 September 1938 to his publisher friend Willy Strecker he wrote: 'There are only two things worth aiming for: good music and a clear conscience, and both of these are now being taken care of.' In response to urgent requests to settle in the USA and because of the difficulty of making a living in Switzerland, he eventually reached New York in February 1940.

During the next few months of that year Hindemith undertook teaching and lecturing jobs arranged by friends – at Buffalo, Cornell and Yale universities, at Wells College (Aurora), and at Tanglewood. In the autumn he was appointed visiting professor of theory of music at Yale. A year later the position was made permanent, and so began an association with Yale that was to last until 1953. His students there were less inclined than those at Berlin to accept his severe demands, and relationships were sometimes tense. He awarded only 12 master's degrees in composition during the whole of his professorship. But he also attracted some of the best talent in the USA and was a powerful influence on American musical life in general, not least in his championship of performances, on authentic instruments, of medieval and Renaissance music. His interest in early music dated back to the 1920s, when he composed and played music for the viola d'amore, and was continued in Berlin where it was stimulated by the work of the musicologist Georg Schünemann. Hindemith's first years in the USA did not provide him with the opportunities for practical music-making he had thrived on in Europe, and in any case his executive abilities were beginning to falter, so the series of concerts of early music he prepared for the Yale Collegium Musicum between 1945 and 1953 served also to provide an outlet for his own performing instincts.

In 1946 Hindemith became an American citizen. The following year he visited Europe for the first time since World War II, giving lectures and appearing as a conductor rather than as an instrumentalist. Although none of his appearances was in Germany he received many offers from there (including ones from Donaueschingen and from the Hochschule at Berlin and at Frankfurt); but he no longer felt that his creative energy depended on living in his homeland. In 1951 he accepted a position at the University of Zurich, and from then until 1953 he divided his time between Zurich and Yale. In 1953 he settled permanently in Switzerland, at Blonay near Vevey. He gave up regular teaching at Zurich in 1955, although he continued, as professor emeritus, to deliver occasional lectures. In his later years he became increasingly attracted to conducting and undertook several major concert tours, including visits to South America and Japan, as well as Europe and the USA. On 15 November 1963 he was taken ill at Blonay and transferred to a hospital in Frankfurt, where he died on 28 December. An autopsy revealed that acute pancreatitis had been the cause of death.

CHAPTER TWO

Works

Hindemith's output after his student years may conveniently be divided into three periods: from 1918 to 1923, when the young composer was exploring a variety of styles; from 1924 until 1933, when he reached a mature neo-Baroque style of considerable harmonic asperity and when, during the latter part of the period, his work with amateurs led to a more lyrical and euphonious mode of expression; and from 1933 until 1963, when he adapted this new and explicitly tonal style to Classical sonata forms and conventional genres. Some works of his late years suggest that he might have developed a fourth period of renewed harmonic asperity and more subtle instrumental colouring; others however contradict this. On balance such ambiguity serves to underline the inventiveness of his music rather than postulate a real change of direction.

I 1918–23

Hindemith's student works are based on traditional models and heavily influenced by Brahms. The first work to break away from conservatory routine was, according to the composer, the lost Piano Quintet of 1917, which developed in continuously interlocking thematic groups 'like a colourful improvisation'. Thereafter he dissociated himself from conservatory teaching and subsequent works seem to have baffled his teachers.

In a letter of May 1917 he wrote: 'I want to write music, not song forms and sonata forms ... I can't talk seriously with anyone because none of them [conservatory teachers] has any ideals left. Their whole art has become far too much craft' (*Hindemith-Jb*, 1972, p.185). Hindemith's creative gifts were thus left to mature on their own.

Superficially his post-1918 music suggests that his method was simply to adopt the colours of various musical factions one by one, retaining or discarding as it suited him. On a deeper level these years reveal a considered philosophy of music, which may best be called international. One episode in particular, dating from his period of military service, seems to have prompted this outlook. During a performance of Debussy's String Quartet news came that Debussy had died; the performance concluded. At that moment, as Hindemith wrote, 'we realized for the first time that music is more than style, technique and the expression of personal feelings. Music stretched beyond political boundaries, national hatreds and the horrors of war. I have never understood so clearly as then what direction music must take' (*Zeugnis in Bildern*, p.8). As far as is known Hindemith did not explain what he meant by the last remark, but the music he subsequently wrote speaks of his determination to break away from German provincialism and inject new blood into the body musical, identifying himself with everything that was fresh in music and with the liberal spirit of hope that prevailed in the early years of the Weimar Republic. Thus the dominant impression given by the early works is of provocative and aggressive novelty, often outrageously offensive to canons of good taste, doubly so because of the facility and appar-

22. Stage design by Ludwig Sievert for the first production of 'Sancta Susanna', Frankfurt, 1922

ent nonchalance with which he composed. A decade was to pass before he had lived down a reputation based on these works.

Hindemith first made his name with music that proclaimed an allegiance with expressionism, notably the one-act operas *Mörder, Hoffnung der Frauen* (1919) and *Sancta Susanna* (1921), the Quartet no.2 (1921) and the ballet *Der Dämon* (1922). In the process his language developed from an amalgam of Puccini, Schreker and Strauss to a more integrated style revealing some affinities with Schoenberg. The texts of the operas are more interesting than their music. Kokoschka's *Mörder, Hoffnung der Frauen*, the first expressionist drama, is written in a symbolic and convulsive manner of almost impenetrable obscurity, though it is clear that the work centres on sexual conflict. By contrast, Stramm's *Sancta Susanna* is an extremely explicit presentation of the sexual fantasies of a young nun. In later years Hindemith regretted he had written these operas, yet they can be said to epitomize the fearless self-assurance of the young composer and his readiness to attempt any workable project. The purely musical aspects of this expressionistic phase are more evident in the Quartet no.2, whose fierce energy and advanced harmonic language sometimes suggest that Hindemith might have drawn close to the aesthetic of the Schoenberg school, though his metrically conceived rhythmic phraseology is foreign to the fluid atonal style of Schoenberg. Hindemith's expressionism belongs to the generally hyper-romantic character of early 20th-century music rather than to any particular aspect of it. Even if this work had represented a wholehearted commitment to expressionism – that is, even if it had avoided passages of regular periodic phrasing and echoes of Ravel (in the finale's second subject) – it would have been surprising if Hindemith had gone any further than he did, for in *Das Nusch-Nuschi* of 1920, the remaining one-act opera of the period, he had already demonstrated that he could venture into a field of chinoiserie charted by Ravel and Stravinsky, and could parody the source of expressionism, *Tristan und Isolde*. The irreverence of *Das Nusch-Nuschi* showed that Hindemith was aware that, in his hands at any rate, expressionism could degenerate into pretentiousness and obscurity.

In the transitional Quartet no.3 of 1922, a work of powerful impulse and sonority, Hindemith still clung, in the first movement, to expressionistic gestures. In the second and third movements he unashamedly though convincingly acknowledged the stimulus of Bartók. In the final two movements, however, he reached towards a more individual style recalling the concertante gestures and disciplined yet expressive lyricism of the Baroque. Although this was in fact the character his music was to adopt, as yet he seems to have been unaware of it, for in his next important work, the *Kammermusik no.1* of the same year, he abruptly aligned himself with Stravinsky and Milhaud, and, by way of a quotation from a contemporary foxtrot and the use of a siren and various other though more traditional noise effects, with the futurists and even dada. In the *Suite '1922'* for piano his interest in the world of jazz and the night club took more concrete form, most of the work's five movements crudely recalling popular dance rhythms. Neither of these works is in a characteristic idiom but, nevertheless, with them Hindemith played out his rejection of expressionism and prepared the ground for the *Kleine Kammermusik*, again of 1922, the first work to reveal the poised hand of a master. If this subtly satirical and witty piece contains relatively few signs of the mature style of a year or two later (the characteristic melodic style of its opening theme is an exception), nevertheless it owes allegiance to no other composer.

Despite their eclecticism and abrupt and apparently inconsequential changes of direction, these early works marked a period not so much of boisterous experiment as of calculated exploration of the territories opened out by Hindemith's seniors. At the same time he gradually discovered his natural style. When most composers made their reputation with orchestral works and full-length operas, Hindemith made his largely with chamber music. His output, if not his style, represented a new attitude, no less significant historically for deriving from the practising musician's natural interest in the genre than for deriving from a considered artistic standpoint. The centre of this output remained the string quartet, but he also revived the solo string sonata, and, taking his cue from *Pierrot lunaire* and *The Soldier's Tale*, approached the idea of chamber music in an extremely flexible way. *Die junge Magd* of 1922, for example, is written for a mixed ensemble of six players accompanying a voice, and the *Kammermusik no.1* for a large ensemble of 12 players including harmonium (later altered to accordion), trumpet and percussion.

Hindemith's underlying seriousness of purpose tended to be obscured by the more spectacular works, but nevertheless it did make unobtrusive appearances in slow movements and in *Die junge Magd*, for example, whose contemplative tone foreshadows *Das Marienleben*. Most of the early works also embody features that were to develop into the typical style of the succeeding neo-classical works. A tendency towards reiterated rhythmic patterns rather than an expressive rubato, a pronounced rhythmic vitality, a highly developed contrapuntal imagination, the use of formal models deriving from the Baroque – all these features can be found in the early works. By the end of the period they had become dominant, so that although such works as the sonata for viola d'amore, the song cycle *Das Marienleben* and the Quartet no.4, all of 1922–3, still employ gestures reminiscent of the 'expressionistic' works, these now take second place to linear counterpoint and clearcut rhythmic patterns. This is particularly apparent in *Das Marienleben*, the most imposing early evidence of the quality of Hindemith's imagination.

Three examples will illustrate Hindemith's development at this time. The intensely chromatic harmony and abundance of melodic appoggiatura in ex.1 provide an

Ex.1 Quartet no. 3 (1922)

Sehr langsame Viertel

Ex.2 Kammermusik no.1 (1922)

instance of his expressionistic vein. In ex.2 he has turned to bitonality, to strident instrumental scoring and to an impish desire to shock, with gestures taken from the cabaret and music hall. By the time of ex.3 he has reached a neo-Baroque style whose sharply modelled lines contain enough diatonic content to make their clashes sound fresh and inevitable. *Das Marienleben* also demonstrates a sure control of large-scale structural organization, which was to stand him in good stead. The 15 songs of the cycle are arranged in four groups, which progress from the lyrical to the dramatic, the poignant and the philosophical, and which employ all manner of Baroque formal schemes, culminating in a final group of ground bass, theme and variations, and chorale fantasia.

Ex.3 'Vom Tode Mariä I' (*Das Marienleben*, 1923)

II 1924–33

It is no accident that the years 1923 and 1924 marked a turning-point in the development of Hindemith's style. By this time the aspirations of the early years of the Weimar Republic had proved illusory and Germany was confronted with inflation, political violence and poverty. In this context the cavalier spirit of the early music was at best inappropriate and at worst morally indefensible. In addition, Hindemith's new position at Donaueschingen demanded a more responsible outlook. He now developed a style explicitly and almost defiantly anti-romantic, which directed attention to the energy in the human soul rather than to its capacity for introversion or self-advertisement. In particular his music drew strength from the gestures and aesthetic of the high Baroque. Most composers in Europe felt the need for greater discipline and objectivity in their music at this period and Hindemith's neo-classicism was exceptional only in that he extended the chromatic vocabulary of his early works (while his contemporaries for the most part favoured more tonal languages) and was unusually explicit in his use of Baroque methods. This had already been apparent in *Das Marienleben* and in the Quartet no.4, whose movements are fugue, chorale prelude, march and passacaglia. In the first important works of this period, the solo concertos comprising the *Kammermusiken nos.2–7* (1924–7), he underlined the point by writing for ensembles that renounce the harmonically orientated balance of 19th-century chamber music and take their stimulus from Bach's Brandenburg Concertos, focussing on solo instruments, especially wind and brass, to give definition to the contrapuntal textures. Each of the six concertos is scored for a different ensemble. In the two other important works of this period, the Concerto for Orchestra of 1925 and the opera *Cardillac* of 1926, Hindemith employed a full orchestra, but one whose wind and brass equal the strings and whose balance is therefore radically different from the 19th-century norm. He used his orchestra like a large-scale chamber ensemble, or reconstituted it to form a number of smaller ensembles often featuring obbligato instruments.

Several other characteristics of these works evoke Baroque models, notably their rhythmic phraseology. Hindemith's use of simply proportioned note values and regular metres, though not necessarily evenly grouped metrical patterns, occasioned the familiar catch phrase of 'motoric rhythms', a workable generalization when applied to the relentless energy of such movements as the first from the *Kammermusik no.5* or the finale from the Concerto for Orchestra, but which ignored the delicacy and subtle manipulation of rhythmic counterpoint in such movements as the finales of the *Kammermusiken nos.3* and *5*, or the third movement of *Kammermusik no.2*. In this last piece Hindemith wittily combined 3/8 and 4/4.

Baroque influence in formal structure is particularly noticeable in his use of ritornellos. These provided scope for clearcut melodic statements, for the use of contrasting instrumental combinations and for the free development of monothematicism, which avoided the dualism of sonata form and its associated leaning towards dramatic as opposed to abstract musical substance. In fact Hindemith did occasionally write sonata movements, as in the finale of *Kammermusik no.4*, but sonata form is here an architectural rather than an evolutionary scheme.

The most individual feature of these works remains their harmony. The rich harmonic vocabulary developed in earlier works was drawn upon, especially in slow movements, but Hindemith rarely organized it into functional progressions, preferring instead to isolate a single harmony in a manner analogous to his use of melodic ritornellos. In any case, this was less characteristic of him than the 'harmony' associated with contrapuntal textures in fast tempos. In such contexts the intervallic tensions deriving from linear counterpoint, and in particular from the release of bass parts from their traditional function as harmonic foundations, produce an astringent level of dissonance that can hardly be said to create harmony at all, an effect heightened by Hindemith's use of strongly triadic chordal progressions at points of structural punctuation. In later life he regarded this aspect of his music as crude because it lacked a theoretical basis. Such passages as that in ex.4 are, however, distinctive, buoyant and invigorating precisely because Hindemith did not tie himself down to functional progressions.

Hindemith's achievement in *Cardillac*, the major work of this period, was to reconcile the Baroque aesthetic with the demands of an expressionistic dramatic subject. Cardillac, a psychopathic goldsmith, murders his customers so that his jewellery shall not fall into the hands of those unable to appreciate it. The opera is laid out in a series of numbers whose structure and gesture is designed to symbolize the dramatic situations. The dramatic atrophy inherent in such a method is counter-

Ex.4 Kammermusik no.4 (1925)

acted by the sequence of numbers, which generates
dynamic impulse by the principle of contrast and by the
use of the unexpected, notably in the final scene of Act
1, which, having prepared for a love-duet, provides
instead a two-part invention for flutes and strings.

In their subject matter and occasional use of jazz
rhythms, the two other operas of this period recall the
irreverent younger composer. *Hin und zurück* was written for the 1927 Baden-Baden Festival, which featured
chamber operas. It is scored for ten players and seven
singers (one of them taking a speaking role and another
remaining mute), and lasts about 12 minutes. Its scenario calls for a reversal of its action from the halfway
point, a device borrowed from contemporary film techniques. The full-length *Neues vom Tage* of 1929 followed the current vogue for 'Zeitopern': operas which,
taking their cue from Krenek's *Jonny spielt auf*, were
topical in subject matter and which included popular
elements within a contemporary idiom. Hindemith's suf-

251

fers from the disparity between its elaborate music and the blunt satire of its tabloid episodes, but the dénouement contains an ironic twist and the work has been unjustly neglected. Both *Neues vom Tage* and *Cardillac* were revised by Hindemith in the 1950s, partly because he then became severely critical of his neo-classical style but principally because he was disturbed by what he took to be the cynicism of the one and the connivance in the philosophy that the artist is above moral law in the other. The revisions however dilute the strength of both works.

Also in 1929 Hindemith wrote *Lehrstück*, one of several politically motivated 'teaching pieces' with texts by Brecht produced at about this time (Hindemith collaborated with Weill on another of them, *Der Lindberghflug*). It is also a unique example of music-theatre, a genre that may be said to have originated in such works as *Pierrot lunaire* and *The Soldier's Tale*, but which in Hindemith's hands became so elastic that there were almost no limits to the diversity of genres that could be absorbed under its heading. *Lehrstück* contains elements of straight opera and oratorio, speaking and acting roles (a clown scene), dance or film, music for offstage brass group, and audience participation. The most important aspect of the work, however, is that all parts are designed to be performed by amateurs. It is Hindemith's most substantial contribution to what became known as *Gebrauchsmusik*.

Hindemith had already given practical and verbal evidence of his conviction that the division between the contemporary composer and the general musical public might be healed if composers wrote with a specific, relevant purpose and use (*Gebrauch*) in mind, and par-

23. *Autograph MS from the final scene, 'Pantomime', of Act 1 of Hindemith's opera 'Cardillac', composed 1926*

ticularly if they encouraged the growth of amateur music. He had written music for films from 1921. From 1926 he wrote for mechanical instruments and for radio; and from 1927 for amateurs, his first such work being the *Spielmusik* for strings, flutes and oboes. His work at the Berlin Hochschule naturally brought him in touch with amateur music. What cemented the connection was a visit to a 'working week' of the musical Jugendbewegung, a movement which at that time had none of the political overtones it was later to acquire, even though its repertory consisted largely of national folk music (with some additional injection of early polyphony). This connection was to prove decisive, for thereby Hindemith encountered German Renaissance popular song, secular and sacred, which was profoundly to influence his compositional style. Whereas his neo-classical music concentrated on continuous, overlapping melodic phrases of instrumental character, he now began to adopt a much more lyrical idiom, whose phraseology, tonal bias and reversion to the principle of melody and accompaniment can be seen to derive from vocal music. The relative simplicity of his amateur music is therefore in no way a capitulation to the demands of musical egalitarianism (as was frequently suggested), but rather an important step in the evolution of a new style.

This amateur music also marked the beginning of a compositional aesthetic he was to maintain for the rest of his life. After 1932 he wrote little that came strictly under the heading of *Gebrauchsmusik* or, as he preferred to call it, *Sing- und Spielmusik*, this in the first instance because he was prevented from so doing in Hitler's Germany. Yet the bulk of his subsequent output remained loyal to his belief that music should be useful and practical, and should not be a vehicle for self-expression. In October 1927, at a lecture in Berlin for a gathering of choral conductors, he said:

The tenuous connection in music today between producers and consumers is to be regretted. The composer today should write only if he knows for what purpose he is writing. The days of composing for the sake of composing are perhaps gone forever. On the other hand the demand for music is so great that it is urgently necessary for composers and users to come to an understanding with each other.

And in the April 1930 issue of *Musik und Gesellschaft* he wrote: 'The performing amateur who seriously concerns himself with musical matters is quite as important a member of our musical life as the professional'. Later in his career Hindemith expressed himself less forcibly but with an idealism that revealed that his beliefs had changed little. In, for example, the 11th chapter of *A Composer's World*, published in 1952, he wrote: 'It is not impossible that out of a tremendous movement of amateur community music a peace movement could spread over the world ... People who make music together cannot be enemies, at least not while the music lasts'.

Apart from *Lehrstück*, the most important Hindemith works of this kind are the short children's opera *Wir bauen eine Stadt* of 1930 and the *Plöner Musiktag* of 1932. This latter provides material for various points of the day at, for example, a summer school of music, ending up with music for a nicely varied evening concert. In all these works Hindemith allowed considerable freedom in the choice of instrumentation.

The first important examples of the new style are the

three *Konzertmusik* works of 1930: for viola and orchestra; for piano, brass and two harps; and for strings and brass. Although their titles already indicate a change of emphasis from 'chamber' works (as in the *Kammermusik* series) to 'concert hall' works fitting to a composer well established in the public arena, they all retain that combination of practicability and adventurousness typical of the younger composer. Their demands can be met by a normal symphony orchestra, yet the resources called for are highly individual. The 'orchestra' of the first of them is weighted in favour of the wind and brass section, the strings consisting of four each of cellos and basses; the remarkable scoring of the second is designed to produce orchestral counterparts to the solo piano's capacity both for hard-edged attack and for delicate resonance; the apparently more conventional layout of the last is deceptively so, for the two groups are presented as opposing blocks and only at the end (where Hindemith, with an abundance of 'blue' notes, characteristically offered his respects to his momentary patron, the Boston SO) do they really cohere. In their formal schemes too they remain distinctive. The first is divertimento-like, the last three of its five movements resembling the chain of dances in a Baroque suite; the other two are each constructed in two extended parts. The crucial difference between these works and their predecessors, however, lies in their revival of a lyric melody, which in its extreme form, as in the opening of the second or in the A major melody in the last part of the third, becomes overtly though distinctively diatonic, and which even in a more chromatic form, as in the opening of the third, is still grounded on diatonic intervals and in any case is constructed in clear phrase groups.

In the oratorio *Das Unaufhörliche* of 1931 Hindemith further developed the implications of his new-found affinity with the principle of melody and accompaniment. Yet in this work something of his natural inventiveness is lost. The process of refinement equipped him to lay out the broad extended paragraph but not to articulate it with the memorable idea. The text of Gottfried Benn argues that man should not stand in the way of the unceasing (*unaufhörlich*) natural law that all things are in a permanent state of transformation. Although this concept has much in common with the matter of the opera *Mathis der Maler*, whose underlying theme is that man (in this case the artist) should not fight against his own nature, the abstract and pessimistic tone of the oratorio is perhaps less typical of the real Hindemith than of his need at that time to seek an alternative to the outlook of Brecht, relations with whom had been severed during the preparations for the festival New Music Berlin 1930. If *Das Unaufhörliche* is one of Hindemith's less durable works it is at least an important step towards the formation of the personal philosophy revealed in *Mathis der Maler*. The *Philharmonisches Konzert* of 1932, a second concerto for orchestra, is a more attractive work because it restores his exuberant and playful invention.

III 1933–63
The idea for an opera about the German painter Matthias Grünewald had been put to Hindemith by Willy Strecker in late summer 1932. Although he at first

rejected it (devoting more attention to another Strecker suggestion, for an opera about Gutenberg), the idea took shape in his mind. Apart from the String Trio no.2 of 1933, his compositional energies during the next three years were concentrated entirely on the opera and its attendant symphony. *Mathis der Maler* is on one level a dramatic allegory about the artist's dilemma in a turbulent society, about Grünewald's decision to renounce his art and commit himself, during the period of the Peasants' Revolt in Germany, to a life of political action, and of his discovery that such action is futile and he must return to his art. It is also a personal testament. In his introduction to the first performance of the opera (Zurich, 1938) Hindemith wrote that Grünewald's experiences had 'shattered his very soul'. Hindemith's own experiences can hardly be said to have done the same, yet the opera's scenes of exaltation and despair and its final scene of resignation depict with disarming frankness the turmoil Hindemith himself lived through and his hard-won solution. He could side neither with the political antipodes nor with the compromising middle ground, and was forced therefore to impose a degree of isolation on himself precisely when his influence and creative ability were at their height, and, ironically, when his musical language was more overtly German than it ever had been. After *Mathis der Maler* something of the excitement generated by his activities was lost. His composing rate began to slow down (this was really apparent however only from 1940) and although his subsequent output occupied about two-thirds of his composing career, it contained few surprises since it was based on the same premises. As a man and as an artist Hindemith had mellowed and come to terms with his essentially conservative temperament. The first scene of *Mathis* contains the words: 'Whatever the deeds blossoming in you, they will flourish only with God's sunlight and only if your roots grow deep into the soil of your people'. Although Mathis later discovers that his 'people', the peasants in the opera, have been corrupted by circumstance, Hindemith himself clearly identified with such sentiments and, in his own libretto, sought to create individual examples of such people: the central figure of Mathis, the idealist Schwalb (leader of the Peasants' Revolt), the pleasure-seeking Archbishop Albrecht (Mathis's patron) who is eventually led to recover his self-respect, the passionate Ursula who is prepared to sell herself for her religious beliefs, the insidious Capito (Albrecht's counsellor), the innocent Regina (Schwalb's daughter, the only figure drawn entirely from Hindemith's imagination), and a number of minor characters – a fair cross-section of German society as Hindemith knew it. The musical consequences of these sentiments grow from his use of German folksongs, secular and sacred, and chorales (his source being Franz Böhme's *Altdeutsches Liederbuch*, Leipzig, 1877). They appear infrequently, but their character strongly influences the style of the opera. In the first scene, for example, Regina sings the folksong 'Es wollt ein Maidlein waschen gehn'. The first half of this is given in ex.5*a*. Mathis responds with a typical theme whose lyricism, phrase length, rhythm and contours echo the simplicity of the song (ex.5*b*).

Like *Cardillac*, *Mathis der Maler* is laid out in separate numbers. But this is concealed, rather than accentuated as in the earlier opera, by carefully engineered recitative or arioso links designed to sustain the continuity of dramatic momentum and variety. The broad sweep of the first scene of the opera, for example, is constructed as follows: ternary-form aria for Mathis followed by three verses of a plainchant hymn (the setting is a monastery); recitative followed by a duet for Mathis and Regina, consisting of the folksong of ex.5a for Regina interspersed with arioso, and another song-like section for both (see ex.5b); recitative and duet for

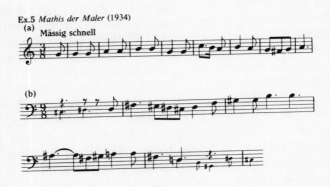

Ex.5 *Mathis der Maler* (1934)

(a) Mässig schnell

(b)

Schwalb and Mathis, consisting of a short ritornello-style aria for Schwalb and a duet proper (in the tradition of the Italian 'vengeance duet') with Mathis; finale in two sections, modelled on the recitative and aria pattern. The scene as a whole is in C♯, this tonality reinforced by its dominant in the first of the duets. Each of the remaining six scenes is also a closed tonal unit, the complete opera moving chromatically from the C♯ to the E of the central scene and then returning in mirror sequence to the beginning.

Although the symphony *Mathis der Maler* may on

superficial examination appear to be a suite of movements extracted from the opera (in fact it was written before work on the opera proper began), Hindemith chose his title deliberately. It is his first symphony, and his first major work employing the tonal and dialectic organization inherent in such a title. Each of the three movements is both a representation of one of the panels from Grünewald's Isenheim altarpiece and a quotation from the opera (the first two movements are taken wholesale, the last is a reworking from scene vi) and, more significantly, part of a large-scale structure. The first movement is a novel re-creation of sonata form, based on the tonality of G and with a strong pull towards C♯. The second is a slow movement, also in sonata form, in which the tonality progresses from the subdominant of G to a clear C♯. The finale is a complex ternary form powerfully asserting the dominance of C♯. The work is thus a coherently argued whole.

In these *Mathis* works Hindemith developed a language that restores the validity of tonality as a structural and expressive tool, while at the same time remaining unmistakably original in its absorption of both traditional and 20th-century harmonic resources. This enabled him to give harmonic function to every bar and to create a euphony that may lack the vivacity and brilliance of earlier works but that gains in depth and range of expression. Of equal significance is Hindemith's recognition of his Classical and Romantic heritage. The programmatic element in the symphony and the new harmonic language confirm that, like so many other composers in the early 1930s, he was seeking a warmer and more humanistic manner.

The careful tonal organization is symptomatic of

Hindemith's growing absorption with the problems of musical syntax and of his search for a contemporary equivalent of the tonal system of earlier centuries. His work in this field grew out of his classes at the Hochschule and eventually resulted in the publication of the first two books of his theoretical work *Unterweisung im Tonsatz* (1937–9). The English title, *The Craft of Musical Composition*, is somewhat misleading, for Hindemith addressed himself largely to the subject of harmony and melody, as one might expect from a German composer. Later he regretted that he had not devoted more attention to other aspects of composition, notably rhythm and form, a deficiency he attempted to make good in the third book of the work. This dissatisfied him, however, for he never published it and in later years intended to write a completely new version of the work as a whole. What remains cannot therefore be regarded as Hindemith's last word on the subject. Nonetheless, two aspects did, for him, take on the character of absolutes, and he was defending them up to his death. The first of these was that the 12 degrees of the chromatic scale are organized, according to acoustical principles, in a precise hierarchy, presenting diminishing degrees of relationship to the first note. The second was that intervals can likewise be organized. Excepting the octave and the tritone, intervals are grouped in pairs, because Hindemith believed them to be invertible and to have roots (marked with an arrow). These two principles are expressed in his series 1 and 2; see ex.6.

Both series thus present relationships and not melodic formulae (though an episode in the finale of the Symphony in E♭ does in fact contain a quotation from series 1). Series 1 enables the composer to chart tonal relationships and tonal centres by assessing the influence of chord roots ('degree progression'). Chords are combinations of intervals, and as such have roots (determined by the lowest, 'best' interval, that nearest the beginning of series 2). Series 2 thus enables the composer to allocate all possible chords to a table of six chord groups, which progress from the triads of group I to the chords of great 'tension' of group IV. (The chords of groups V and VI are ambiguous and transitional.) By analysis of such harmonic fluctuation the composer can thereby determine the smoothness or otherwise of a progression. Series 2 also distinguishes between intervals primarily harmonic (near the beginning) and those primarily melodic in character.

Ex.6

Hindemith believed the force of tonality, as exemplified at the beginning of each series, to be so strong as to be inescapable, except by 'sheer perversity' (by which he meant atonality): 'Music, as long as it exists, will take its departure from the major triad and return to it'. In addition, he believed that his discoveries, especially his harmonic discoveries, were of fundamental importance to music as a whole, that he had provided the basis for a *lingua franca* by which the merits of a composer could be judged in relation to accepted criteria. 15 years after publishing *Unterweisung im Tonsatz* he could still write: 'Music has now entered the phase of its life which corresponds with the natural permanent state of poetry' (*A Composer's World*, 'Technique and Style'). His own music, at least up to the works of his final years, testifies to the position he felt morally obliged to uphold: that of defender of tradition and 'lasting values'. Had Europe not succumbed to fascism, he might well have shed his emotional reticence and again played the pioneer role of his earlier years. As it was, Hindemith seemed determined not to subjectify his predicament as an artist in exile. The warmth and passion that entered his music with *Mathis der Maler* tended thereafter to keep their distance: Hindemith objectified his introversion and spent the rest of his career consolidating the musical aspects of the language he had fashioned, while raising to the level of a principle the initially pragmatic aims of his amateur music.

In the preface to *Der Schwanendreher* of 1935, his third viola concerto, he provided a touching though perhaps unwitting self-portrait: 'A minstrel visits a happy company of people and plays for them the music he has brought from far away. According to his fancy and ability he develops and decorates the old tunes with preludes and fantasias'. This work is based on folksongs, and if Hindemith could not play with the happy company he imagined, he could remind them of the temperament enshrined in their songs by playing for them, or at least making himself understood through the printed page. He could also write music suitable for domestic music-making. The first important work to follow *Mathis der Maler* was the Violin Sonata in E, itself the first of an extensive corpus of sonatas for almost every instrument. In his five remaining years in Europe he wrote 16 of the 25. Not all are technically as easy as this sonata, but all remain focussed on the accomplishments of a gifted amateur. He could also underline this attitude in concert works, by again writing for conventional genres and resources, and demonstrating that in his hands traditional procedures and attitudes could stimulate an inexhaustible flow of invention. In 1939 he wrote his Violin Concerto, the first of eight concertos which build on the tradition of the genre in the 19th century while not foresaking the concertante approach of the 18th. In 1940 he wrote the Symphony in E♭, the first of five symphonies of similar purpose. In all this output one can detect a vision of a music that might synthesize the inheritance of tonal music from the Middle Ages to the 20th century and assert the archetypal nature of forms, procedures and even themes from past eras by absorbing them within the all-embracing style of Hindemith's own language. In this sense the music is valedictory, an impression strengthened by the frequent use of downward contours in thematic material and of the chaconne bass in accompaniments which give the harmonic progressions a leaning towards the flat side of the tonics.

The position held by Hindemith's harmonic language in his post-*Mathis* music was crucial. It enabled him to exercise a firm control over tonal progression and thereby consciously to organize tonal structure. The strength of this language is most apparent in his cadences, examples of which are given in ex.7. His chief intention was to give dynamic logic to his progressions. Only late in his life was he able to allow the relation between function and timbre to lean in the direction of the latter, and this but rarely (as at the end of Act 5

Ex.7

(a) Suite *Nobilissima visione* (1938)

(b) *Ludus tonalis* (1942)

(c) *Die Harmonie der Welt* (1957)

scene i of *Die Harmonie der Welt* or in the third movement of the Organ Concerto of 1962). His orchestration
is thus designed to clarify harmonic and linear function,
again with the proviso that in later years he developed a
more sensitive feel for colour. The principle of melody
with harmonic accompaniment remained fundamental.
This would have been less obvious had the structure of
his melodies not been so similar. (Hindemith seems to
have been aware of this difficulty, for he also evolved a

novel form of melodic construction by which details were determined by an underlying verbal text, as in the ballet *Hérodiade* of 1944.) Few of his melodies cannot be analysed as variants on the four-phrase sentence deriving from their ancestry in the folksong or chorale. Hindemith's gain here is that the ear is obliged to concentrate on the poise and beauty of the melodies *per se* (see ex.8), which in turn directs attention more towards their quality as paradigms of proportion than towards their emotional or impressionistic content. The danger is not only that such melodies become routine, but also that a failing of the aural imagination can jeopardize the quality of the melody as a whole, as when a component part is filled in with perfunctory invention because Hindemith's anxiety to reach a focal tonal point overrides his need to clothe the process with significant material. The fourth phrase (see brackets) of ex.8*c* is a case in point.

Formal processes therefore assume the greatest importance as symbols of a transcendent balance and proportion Hindemith believed to be music's heritage. He also asked the listener to hear music in this way, and he eased the process by favouring schemes which offer the listener a chance to savour the skill with which the composer acquits himself, such as fugues, variations, passacaglias, cantus firmus movements, and his idiosyncratic habit of combining themes and tempos of dissimilar character. He used the sonata principle less to reconcile dramatic conflict than to provide a model for the elaboration of multi-thematic premises. The opening theme of the Cello Concerto of 1940 (ex.8*b*), for example, contains on its own enough material for a

Ex.8

(a) Cello Concerto (1940) 2nd movement

Andante con moto

(b) Cello Concerto (1940) 1st movement

Allegro moderato

movement, but Hindemith's sense of phrase structure is so ingrained that the character of subsidiary motifs is subsumed by the force of the linear impulse.

This reluctance to give free rein to the individual idea (unless this is structurally necessary, as in a develop-

(c) Requiem (1946)
Slow, solemn

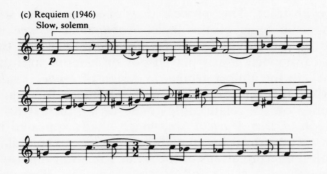

ment section) is characteristic of Hindemith's compositional identity. It accounts for the relentless yet rather prosaic nature of his rhythmic invention, which abounds in the functional and conscious use of rhythmic modulation, rhythmic differentiation and rhythmic counterpoint rather in the manner of Brahms, yet which in essence is unable to escape from its roots in the heavy metric accents of the march or chorale. Hindemith attempted to overcome this dilemma by theorizing about the nature of rhythm; but he never published his '19th exercise' for *Unterweisung im Tonsatz* and was obviously not satisfied with it (it appears in Briner, pp.323ff).

Because Hindemith's later music reflects a fixed artistic purpose, it is not easy to perceive important changes of direction, or to subdivide his output into neat groups. It can be said, however, that in the post-*Mathis* works of the 1930s and 1940s his style became more diatonic and explicitly tonal, concentrating on the mildly dissonant chords of his group III and on harmonic progressions that softened the emotional tension of the chromatic content. At the same time his music became more abstract. This process developed from the still humanistic basis of the ballet *Nobilissima visione* of 1938 (a counterpart to *Mathis der Maler* in which the idea of renunciation is played out on a more spiritual plane), to the almost geometric design of *Ludus tonalis* of 1942 (a cycle of fugues and interludes for piano founded on the relationships of series 1), the extensive revision of *Das Marienleben*, which finally appeared in 1948 accompanied by an elaborate preface expounding mystical interpretations of the series 1 relationships, and ultimately to his last full-length opera, *Die Harmonie der Welt*, which although set as a series of dramatic tableaux is an overtly mystical work dramatizing the attempts of the Renaissance astronomer Kepler to discover the harmony of the spheres. In all these works Hindemith can be seen struggling with the idea that his theoretical concepts might give rise to music of such purity that it could provide a gateway into the secrets of the universe. In this connection the final scene of the opera (which is presented intact, though without voices, in the companion symphony *Die Harmonie der Welt*) is an illuminating indication of his position. Although earlier parts of both opera and symphony contain passages of much greater harmonic acerbity than had been found in Hindemith's previous work, this final scene, a passacaglia, eventually opens out on a brilliant triad of E major, a symbol of the putative 'harmony' that may be said to stretch beyond the formal ending of the work.

Hindemith's first ideas for the opera dated from the late 1930s (the sonatas written during that period he regarded as a technical preparation) and although he delayed so much that he eventually composed under pressure, it remained the culmination of his output. That

24. Paul Hindemith, 1957

Hindemith should have identified so closely with Kepler is paradoxical, for his own Protestant instinct to invoke the sanction of ultimate authority (his two series) matches uneasily the boundless aspirations of so typically Renaissance a figure. In the music of his final years Hindemith tried to break out into new territory, by exploring an increasingly dissonant harmonic vocabulary and by attempting some form of reconciliation with Schoenberg (whom he had vehemently opposed during the older man's lifetime). In the Tuba Sonata of 1955 he used a 12-note theme and in the finale of the

Pittsburgh Symphony of 1958 his basic material is an eight-note series of uncharacterized pitches. Yet his faith in the triad never wavered. Certain passages in his late works consist exclusively of triads, the sextet from the one-act opera *The Long Christmas Dinner* of 1960, for example, and the Canzonetta movement of his Organ Concerto of 1962. And in the final bars of his last work, the Mass of 1963, he was still able to draw great beauty from the harmonic language he had espoused for the last 30 years of his life.

Despite its originality, or perhaps because of its consistency, Hindemith's music has exerted no significant influence on composers of later generations. Yet its authentic sound and consummate technique assures its status both as a body of finished works of art of unassailable integrity and as a fitting monument to the neo-classical ideals of the 20th century.

Edition: *P. Hindemith: Sämtliche Werke*, ed. K. von Fischer and L. Finscher (Mainz, 1975–)

Numbers in the right-hand column denote references in the text.

DRAMATIC

op.	Title	Genre (acts, libretto/scenario)	Composition	First performance	
12	Mörder, Hoffnung der Frauen	opera (1, O. Kokoschka)	1919	cond. F. Busch, Stuttgart, Landestheater, 4 June 1921	231, 241
20	Das Nusch-Nuschi	play for Burmese marionettes (1, F. Blei)	1920	cond. Busch, Stuttgart, Landestheater, 4 June 1921	231, 242, 243
		dance suite, orch	1920		*241*, 241
21	Sancta Susanna	opera (1, A. Stramm)	1921	cond. L. Rottenberg, Frankfurt, Opernhaus, 26 March 1922	
—	In Sturm und Eis [written under the pseudonym Paul Merano]	film score, for A. Fank: Im Kampf um den Berge	1921, unpubd		
—	Tuttifäntchen	Christmas fairy tale (3 scenes, H. Michel, F. Becker)	1922	cond. W. Beck, Darmstadt, 13 Dec 1922	
28	Der Dämon	dance-pantomime (2 scenes, M. Krell)	1922	cond. J. Rosenstock, Darmstadt, Landestheater, 1 Dec 1923	242
		concert suite	1923		
39	Cardillac	opera (3, F. Lion, after Hoffmann: Das Fräulein von Scuderi) (rev. lib Hindemith after Lion)	1925–6, rev. 1952	cond. Busch, Dresden, Staatsoper, 9 Nov 1926; rev. version cond. V. Reinshagen, Zurich, Stadttheater, 20 June 1952	249, 250–51, 252, *253*, 259
40/2	Das triadische Ballett	abstract ballet (O. Schlemmer), mechanical org	1926, unpubd	Donaueschingen, 25 July 1926	
45a	Hin und zurück	sketch (1, M. Schiffer)	1927	cond. E. Mehlich, Baden-Baden, 15 July 1927	251
44/1	Felix der Kater im Zirkus	film score, mechanical org	1927, unpubd	Donaueschingen, July 1927	
—	Vormittagsspuk, lost	film score (H. Richter), mechanical pf	1928, unpubd		
—	Neues vom Tage	comic opera (3 pts., Schiffer)	1928–9, rev. 1953	cond. O. Klemperer, Berlin, Kroll, 8 June 1929	251, 252
		concert version of ov.	1930		

op.	Title	Genre (acts, libretto/scenario)	Composition	First performance	
—	Lehrstück	[music theatre work] (Brecht)	1929	cond. A. Dressel, Baden-Baden, 28 July 1929	252, 255
—	Der Lindberghflug	radio play (Brecht) [1st version, with music by Hindemith and Weill]	1929	cond. H. Scherchen, Baden-Baden, 28 July 1929	252
—	Wir bauen eine Stadt	play for children (R. Seitz)	1930	Berlin, 21 June 1930	235, 255
—	Sabinchen	radio play	1930, unpubd		
—	Clermont de Fouet	film score, str trio	1931, unpubd		
—	Mathis der Maler	opera (7 scenes, Hindemith)	1934–5	cond. R. F. Denzler, Zurich, Stadttheater, 28 May 1938	257–61, 264, 269, 270
—	Nobilissima visione	dance legend (6 scenes, Massine)	1938, reorchd 1939	cond. Hindemith, London, Covent Garden, 21 July 1938	270
—	Thema mit vier Variationen (Die vier Temperamente)	suite ballet	1938 1940	Ballet Society (G. Balanchin), New York, 20 Nov 1946	
—	Hérodiade	ballet (orch recitation after S. Mallarmé)	1944	Martha Graham, Washington, DC, Library of Congress, 30 Oct 1944	267
—	Die Harmonie der Welt	opera (5, Hindemith)	1956–7	cond. Hindemith, Munich, Prinzregententheater, 11 Aug 1957	265–6, 270
—	The Long Christmas Dinner	opera (1, T. Wilder)	1960	cond. Hindemith, Mannheim, Nationaltheater, 17 Dec 1961	272

ORCHESTRAL

3	Cello Concerto, Eb, 1916, unpubd	
4	Lustige Sinfonietta, small orch, 1916, unpubd	
24/1	Kammermusik no.1, small orch, 1922	231, 243, 244, 246, 256
29	Klaviermusik mit Orchester, pf left hand, orch, 1923, unpubd	
36/1	Kammermusik no.2 (Pf Conc.), pf, 12 insts, 1924	248, 249
36/2	Kammermusik no.3 (Vc Conc.), vc, 10 insts, 1925	248, 249
36/3	Kammermusik no.4 (Vn Conc.), vn, large chamber orch, 1925	248, 250
36/4	Kammermusik no.5 (Va Conc.), va, large chamber orch, 1927	235, 248, 249
38	Concerto for Orchestra, 1925	249
41	Konzertmusik, wind, 1926	
46/1	Kammermusik no.6, va d'amore, chamber orch, 1927, 2 versions	248
49	Konzertmusik, pf, brass, 2 harps, 1930	256
50	Konzertmusik, brass, str, 1930	256
—	Konzertstück, trautonium, str, 1931, unpubd	257
—	Philharmonisches Konzert, variations, 1932	236, 258, 260–264, 269
—	Mathis der Maler, sym, 1934	264
—	Der Schwanendreher, conc. after folksongs, va, small orch, 1935	264
—	Trauermusik, va/vn/vc, str, 1936	235
—	Symphonische Tänze, 1937	
—	Violin Concerto, 1939	265
—	Cello Concerto, 1940	267–8
—	Symphony, Eb, 1940	262, 265
—	Cupid and Psyche, ballet ov, 1943	

Piano Concerto, 1945 —

Symphonia serena, 1946 —

Clarinet Concerto, 1947 —

Horn Concerto, 1949 —

Concerto, fl, ob, cl, bn, harp, orch, 1949 —

Concerto, tpt, bn, str, 1949 —

Sinfonietta, E, 1950 —

Symphony, Bb, concert band, 1951 —

Symphony 'Die Harmonie der Welt', 1951 — 270

Pittsburgh Symphony, 1958 — 272

Marsch über den alten 'Schweizerton', 1960 —

Organ Concerto, 1962 — 266, 272

CHORAL

(accompanied)

Lügenlied, 3vv, wind and str ad lib, 1928 —

Das Unaufhörliche (G. Benn), oratorio, S, T, Bar, B, chorus, children's chorus, orch, 1931 — 257

Ausflugskantate, 1v, chorus, 4 cl, 1934, unpubd —

Old Irish Air, SATB, pf/(harp, str), 1940 —

A Song of Music (G. Tyler), SSA, pf/str, 1941 —

When Lilacs Last in the Door-yard Bloom'd: Requiem for those we Love (Whitman), Mez, Bar, chorus, orch, 1946 — 267

Apparebit repentina dies (anon., c700), chorus, 10 brass, 1947 —

Ite, angeli veloces (Claudel), cantata in 3 pts.: Chant de triomphe du roi David (Ps.xvii), A, T, chorus, audience, orch, wind orch, 1955; Custos quid de nocte, T, chorus, orch, 1955; Cantique de l'espérance, Mez, chorus, audience, orch, wind orch, 1953 —

Mainzer Umzug (Zuckmayer), S, T, Bar, chorus, orch, 1962 —

(unaccompanied)

Lieder nach alten Texten, 1923: Vom Hausregiment (Luther), SSATBarB; Frauenklage (Burggraf zu Regensburg), SSATB; Art lässt nicht von Art (Spervogel), SATB; Der Liebe Schrein (H. von Morungen), SSATB; Heimliches Glück (Reinmar), SSATBarB; Landsknechtstrinklied, SSATBarB — 33

Zwei Lieder, 3vv, 1927, unpubd —

Spruch eines Fahrenden (14th century), female/children's vv, 1928 —

Über das Frühjahr (Brecht), TTBB, 1929 —

Eine lichte Mitternacht (Whitman, trans. Schlaf), TTBB, 1929 —

Chorlieder für Knaben, 1930 —

Du musst dir alles geben (Benn), TTBB, 1930 —

Fürst Kraft (Benn), TTBB, 1930 —

Vision des Mannes (Benn), TTBB, 1930 —

Der Tod (Hölderlin), TTBB, 1932 —

Wahre Liebe (H. von Veldecke), SSATB, 1936 —

Five Songs on Old Texts, SSATB, 1936: True Love, Lady's Lament, Of Household Rule, Trooper's Drinking Song, The Devil a Monk would be [no.1 Eng. version of Wahre Liebe, nos.2-5 after op.33] —

Six Chansons (Rilke), SATB, 1939: La biche, Un cygne, Puisque tout passe, Printemps, En hiver, Verger —

Drei Chöre, TTBB, 1939: Das verfluchte Geld (old Ger.), Nun da der Tag des Tages müde ward (Nietzsche), Die Stiefmutter (old Ger.) —

Erster Schnee (Keller), TTBB, 1939 —

Variationen über ein altes Tanzlied, TTBB, 1939 —

The Demon of the Gibbet (Das Galgenritt) (F. J. O'Brien, trans. Hindemith), TBB, 1939 —

Vier Kanons, female vv: Sine musica nulla disciplina (Hrabanus Maurus), 3vv, 1946; Musica divinas laudes (old proverb), 3vv, 1949; Hie kann nit sein ein böser Mut (old proverb), 3vv, 1928; Wer sich die Musik erkiest (Luther), 2vv, 1928 —

Kanon: Musica divina laudes (old proverb), children's/female vv, 1949 —

Zwölf Madrigale (J. Weinheber), SSATB, 1958: Mitwelt; Eines Narren, eines Künstlers Leben; Tauche deine Furcht in schwarzen Wein; Trink aus!; An eine Tote; Frühling; An einen Schmetterling; Judaskuss; Magisches Rezept; Es bleibt wohl, was gesagt wird; Kraft fand zu Form; Du Zweifel an dem Sinn der Welt —

Mass, 1963 — 272

SOLO VOCAL

(with orchestra or ensemble)

Drei Gesänge, S, orch, 1917, unpubd — 9

Melancholie (Morgenstern), 4 songs, Mez, str qt, 1919, unpubd — 13

No.	Work	Page refs
23a	Des Todes Tod (E. Reinacher), female v, 2 va, 2 vc, 1922: Gesicht von Tod und Elend, Gottes Tod, Des Todes Tod	
23/2	Die junge Magd (Trakl) A, fl, cl, str qt, 1922: Oft am Brunnen, Stille schafft sie in der Kammer, Nächtens übern kahlen Anger, In der Schmiede dröhnt der Hammer, Schmächtig hingestreckt im Bette, Abends schweben blutige Linnen	244
35	Die Serenaden, little cantata, S, ob, va, vc, 1925: Barcarole (A. Licht); An Phyllis (J. L. W. Gleim); Corrente; Nur Mut (L. Tieck); Duet, va, vc; Der Abend (Eichendorff); Der Wurm am Meer (J. W. Meinhold); Trio, ob, va, vc; Gute Nacht (S. A. Mahlmann)	
—	Das Marienleben (Rilke), S, orch, 1938–59: Geburt Mariä, Argwohn Josephs, Geburt Christi, Rast auf der Flucht nach Ägypten, Vor der Passion, Vom Tode Mariä III	

(with piano)

No.	Work	Page refs
5	Lieder in Aargauer Mundart, 1916, unpubd	
14	Drei Hymnen von Walt Whitman, Bar, pf, 1919, unpubd	
18	Acht Lieder, S, pf, 1920: Die trunkene Tänzerin (C. Bock); Wie Sankt Franciscus schweb' ich in der Luft (Morgenstern); Auf der Treppe sitzen meine Öhrchen (Morgenstern); Vor dir schein' ich aufgewacht (Morgenstern); Du machst mich traurig – hör' (Lasker-Schüler); Durch die abendlichen Gärten (H. Schilling); Trompeten (Trakl)	244, 245, 246–7, 248, 270
27	Das Marienleben (Rilke), S, pf, 1922–3: Geburt Mariä, Die Darstellung Mariä im Tempel, Mariä Verkündigung, Mariä Heimsuchung, Argwohn Josephs, Verkündigung über die Hirten, Geburt Christi, Rast auf der Flucht nach Ägypten, Vor der Hochzeit zu Kana, Von der Passion, Pietà, Stillung Mariä mit dem Auferstandenen, Vom Tode Mariä I, Vom Tode Mariä II, Vom Tode Mariä III; rev. 1936–48; 6 nos. orchd 1938–59	
—	Vier Lieder (Claudius), S, pf; Vier Lieder (Rückert); Vier Lieder (Novalis); Drei Lieder (W. Busch); all 1933, unpubd	
—	Sechs Lieder (Hölderlin), T, pf: An die Parzen, 1935; Sonnenuntergang, 1935; Ehmals und jetzt, 1935; Des Morgens, 1935; Fragment, 1933; Abendphantasie, 1933	
—	Vier Lieder (Silesius), 1935; Zwei Lieder (C. Brentano), 1936; Das Köhlerweib (Keller), 1936; In principio erat Verbum, motet, S, pf, 1941; 18 lieder (Ger., Fr., Lat.), S, pf, 1942; Drei Lieder (Fr.), S, pf, 1944; all unpubd	
—	La belle dame sans merci (Keats), 1942	244
—	Nine English Songs, S/Mez, pf: Echo (Moore), 1942; Envoy (Thompson), 1942; The Moon (Shelley), 1942; On a Fly Drinking out of his Cup (Oldys), 1942; On Hearing 'The Last Rose of Summer' (C. Wolfe), 1942; The Wild Flower's Song (Blake), 1942; The Whistlin' Thief (S. Lover), 1942; Sing on there in the Swamp (Whitman), 1943; To Music, to Becalm his Fever (Herrick), 1944	
—	Two Songs (O. Cox), S/T, pf, 1955: Image, Beauty touch me	
—	Dreizehn Motetten, S/T, pf: Exiit edictum, 1960; Pastores loquebantur, 1944; Dicebat Jesus scribis et pharisaeis, 1959; Dixit Jesus Petro, 1959; Angelus Domini apparuit, 1958; Erat Joseph et Maria, 1959; Defuncto Herode, 1958; Cum natus esset, 1941; Cum factus esset Jesus, 1959; Vidit Joannes Jesum, 1959; Nuptiae factae sunt, 1944; Cum descendisset Jesus, 1960; Ascendente Jesu in naviculam, 1943	

CHAMBER AND INSTRUMENTAL
(for three or more instruments)

No.	Work	Page refs
1	Andante and Scherzo, cl, hn, pf, 1914, unpubd, lost	
2	String Quartet, c, 1915, unpubd	
7	Piano Quintet, e, 1917, unpubd, lost	231, 239
10	String Quartet no.1, f, 1919	231
16	String Quartet no.2, C, 1921	231, 234, 241, 242
22	String Quartet no.3, 1922	243, 245
24/2	Kleine Kammermusik, wind qnt, 1922	235, 243
30	Clarinet Quintet, 1923, 2 versions	
32	String Quartet no.4, 1923	245, 248
34	String Trio no.1, 1924	
—	Three Pieces, cl, tpt, pf, vn, db, 1925	
—	Rondo, 3 gui, 1925	
—	Zwei kleine Trios, fl, cl, db, 1927, unpubd	
47	Trio, pf, va, heckelphone/t sax, 1928	
—	Triosatz, 3 gui, 1930, unpubd	
—	String Trio no.2, 1933	258
—	Unterhaltungsmusik, 3 cl, 1934, unpubd	

— Folksong arrangements, cl, str qnt, 1936, unpubd, lost
— Quartet, cl, pf trio, 1938
— String Quartet no.5, Eb, 1943
— String Quartet no.6, 1945
— Septet, fl, ob, cl, b cl, bn, hn, tpt, 1948
— Sonata, 4 hn, 1952
— Octet, cl, bn, hn, vn, 2 va, vc, db, 1957–8

(for one or two instruments)

8   Three Pieces, vc, pf, 1917
11 1 Sonata, Eb, vn, pf, 1918; 2 Sonata, D, vn, pf, 1918; 3 Sonata, vc, pf, 1919; 4 Sonata, F, va, pf, 1919; 5 Sonata, va, 1919; 6 Sonata, g, vn, 1917, unpubd, lost   231
25 1 Sonata, va, 1922; 2 Kleine Sonate, va d'amore, pf, 1922; 3 Sonata, vc, 1922; 4 Sonata, va, pf, 1922
31 1 Sonata, vn, 1924; 2 Sonata, vn, 1924, 3 Kanonische Sonatine, 2 fl, 1923; 4 Sonata, va, 1923, unpubd   235
— Eight Pieces, fl, 1927; other pieces unpubd
— Nine Pieces, cl, db, 1927, unpubd
— Zwei kanonische Duette, 2 vn, 1929
— Vierzehn leichte Stücke, 2 vn, 1931
— Konzertstück, 2 a sax, 1933
— Duet, va, vc, 1934
— Sonata, E, vn, pf, 1935
— Sonata, fl, pf, 1936
— Sonata, va, 1937, unpubd
— Drei leichte Stücke, vc, pf, 1938   236, 264–5
— Sonata, bn, pf, 1938
— Sonata, ob, pf, 1938
— Meditation [from Nobilissima visione], vn/va/vc, pf, 1938
— Sonata, hn, pf, 1939
— Sonata, tpt, pf, 1939
— Sonata, harp, 1939
— Sonata, C, vn, pf, 1939
— Sonata, C, va, pf, 1939
— Sonata, cl, pf, 1939
— Sonatas nos.1–3, org, 1937, 1937, 1940
— A Frog he went a-courting, variations, vc, pf, 1941
— Sonata, eng hn, pf, 1941

— Sonata, trbn, pf, 1941
— Pieces, bn, vc, 1941, unpubd
— Kleine Sonate, vc, pf, 1942, unpubd
— Echo, fl, pf, 1942
— Sonata, a hn/hn/a sax, pf, 1943
— Ludus minor, cl, vc, 1944, unpubd
— Sonata, vc, pf, 1948
— Sonata, db, pf, 1949
— Sonata, b tuba, pf, 1955   271

(for mechanical and electronic instruments)

40 Musik für mechanische Instrumente: 1 Toccata für mechanisches Klavier [Welte-Mignon], 1926, unpubd; 2 Das triadische Ballett für mechanische Orgel, 1926, unpubd

44 Musik für mechanische Instrumente: 1 Musik zum Film 'Felix der Kater im Zirkus', 1927, unpubd; 2 Suite für Orgel [arr. of pt.1 of Das triadische Ballett]
— Filmmusik Vormittagsspuk, mechanical pf, 1928, unpubd, lost
— Four Pieces, 3 trautoniums, 1930, unpubd
— Experiments on 2 gramophone discs, 1930, lost
— Konzertstück, trautonium, str, 1931, unpubd
— Langsames Stück and Rondo, trautonium, 1935, unpubd, lost
[The entries for opp.40 and 44 above are transcribed from Hindemith's private catalogue of his music. Since he never intended to publish his mechanical music, his subsequent use of op.44 for Schulwerk (see below) would not have caused confusion.]

PIANO

6   Sieben Walzer, 4 hands, 1916, unpubd
15 In einer Nacht, 1919, unpubd
17 Sonata, 1920, unpubd, lost
19 Tanzstücke, 1922
26 Suite '1922', 1922
37 Klaviermusik, 1925–7   243
— Zwei kleine Klavierstücke, 1934, unfinished, unpubd
— Sonatas nos.1–3, A, G, Bb, 1936
— Sonata, 4 hands, 1938
— Ludus tonalis, 1942   270
— Sonata, 2 pf, 1942

SING- UND SPIELMUSIK

43 | 1 Spielmusik, 2 fl, 2 ob, str orch, 1927; 2 Lieder für Singkreise, 3vv, 1927: Ein jedes Band (A. von Platen); O Herr, gib jedem seinen eignen Tod (Rilke); Man weiss oft grade denn am meisten (Claudius); Was meinst du, Kunz, wie gross die Sonne sei (Claudius) — 254

44 | Schulwerk für Instrumental-Zusammenspiel, 1927: I 9 Pieces, 2 vn; II 8 Canons, 2 vn, vn/va; III 8 Pieces, str qt, db; IV 5 Pieces, str orch — 254

45 | Sing- und Spielmusiken für Liebhaber und Musikfreunde:
I Frau Musica (Luther), Mez, Bar, chorus, str orch, wind ad lib, 1928, rev. 1943
II 8 Canons, 2vv, str qt ad lib, 1928: Hie kann nit sein ein böser Mut (old proverb); Wer sich die Musik erkiest (Luther); Die wir dem Licht in Liebe dienen (R. Goering); Auf a folgt b (Morgenstern); Niemals wieder will ich eines Menschen Antlitz verlachen (Werfel); Das weiss ich und hab' ich erlebt (J. Kneip); Mund und Augen wissen ihre Pflicht (H. Claudius); Erde, die uns dies gebracht (Morgenstern)
III Ein Jäger aus Kurpfalz, der reitet durch den grünen Wald, wind, str, 1928
IV Kleine Klaviermusik, 1929
V Martinslied (J. Olorinus), 1v/chorus, 3 str/wind, 1929
— Wir bauen eine Stadt (play for children, R. Seitz), 1930

— | Plöner Musiktag, 1932: — 255
A Morgenmusik, 2 tpt/flügelhn, 2 rbrn/hn, tuba ad lib
B Tafelmusik, fl, tpt/cl, str
C Kantate (Agricola), 1v, speaker, children's chorus, str orch, wind and perc ad lib

D Abendkonzert: Einleitungsstück, orch; Flötensolo mit Streichern; 2 Duets, vn, cl; Variations, cl, str; Trio, 3 rec; Quodlibet, orch — 254, 254
— Nine little songs for an American school songbook, 1938, unpubd

EDITIONS, ARRANGEMENTS ETC

Works by A. Vivaldi, 1932; H. Biber, 1933; cadenzas for concertos by Mozart, 1933–4; A. Ariosti, 1935; C. Monteverdi, Orfeo, 1943; Suite französischer Tänze, after Gervaise, du Tertre, 1958
Principal publisher: Schott
For fuller list of unpubd works see Briner, pp.369ff

WRITINGS

Unterweisung im Tonsatz, i: Theoretischer Teil (Mainz, 1937, rev. 2/ 1940; Eng. trans, 1942, rev. 2/1948) — 234, 262–3, 269
Unterweisung im Tonsatz, ii: Übungsbuch für den zweistimmigen Satz (Mainz, 1939; Eng. trans, 1941) — 234, 262–3, 269
A Concentrated Course in Traditional Harmony, i (New York, 1943, 2/1949; Ger. trans., 1949)
Elementary Training for Musicians (New York, 1946, rev. 2/1949)
A Concentrated Course in Traditional Harmony, ii: Exercises for Advanced Students (New York, 1948, 2/1953; Ger. trans., 1949)
A Composer's World (Cambridge, Mass., 1952; Ger. trans., 1959) [Ger. edn. incl. addn to chap.5; Eng. trans. in JMT, v (1961), 109] — 255, 264
Johann Sebastian Bach: ein verpflichtendes Erbe (Frankfurt, 1953; Eng. trans., 1952)
Unterweisung im Tonsatz, iii: Der dreistimmige Satz (Mainz, 1970)

BIBLIOGRAPHY

CATALOGUES AND BIBLIOGRAPHIES

K. Stone: *Paul Hindemith: Catalogue of his Works and Recordings* (London, 1954)

E. Westphal: *Paul Hindemith: eine Bibliographie des In- und Auslandes seit 1922* (Cologne, 1957)

O. Büthe: *Paul Hindemith: Emigration und Rückkehr nach Europa* (Frankfurt, 1965) [exhibition catalogue]

Paul Hindemith: Werkverzeichnis (Mainz, 1969)

H. Rösner: *Paul Hindemith: Katalog seiner Werke, Diskographie, Bibliographie, Einführung in das Schaffen* (Frankfurt, 1970)

E. Kraus: 'Bibliographie: Paul Hindemith', *Musik und Bildung*, iii (1971), 249

H. Rösner: 'Zur Hindemith-Bibliographie', *Hindemith-Jb*, i (1971), 161–95

O. Zickenheimer: 'Hindemith-Bibliographie 1971–73', *Hindemith-Jb*, iii (1974), 155–94

A. Laubenthal: 'Hindemith-Bibliographie 1974–8', *Hindemith-Jb*, vii (1978), 229

MONOGRAPHS AND COLLECTIONS OF ESSAYS

H. Strobel: *Paul Hindemith* (Mainz, 1928, 3/1948)

Paul Hindemith: Testimony in Pictures/Zeugnis in Bildern (Mainz, 1961)

H. L. Schilling: *Paul Hindemiths Cardillac* (Würzburg, 1962)

Paul Hindemith: Die letzten Jahre (Mainz and Zurich, 1965)

I. Kemp: *Hindemith* (London, 1970)

A. Briner: *Paul Hindemith* (Zurich and Mainz, 1971)

Musikrevy, xxvi (1971) [Hindemith issue]

Hindemith-Jb, i– (1971–)

Revue musicale de Suisse romande, xxvi (1973) [Hindemith issue]

G. Skelton: *Paul Hindemith: the Man behind the Music* (London, 1975)

I. Prudnikova, ed.: *Paul' Khindemit: stat'i i materialï* (Moscow, 1979)

G. Schubert: *Hindemith* (Hamburg, 1981)

D. Rexroth: *Paul Hindemith Briefe* (Frankfurt, 1982)

OTHER LITERATURE

W. Hymanson: 'Hindemith's Variations', *MR*, xiii (1952), 20

N. Cazden: 'Hindemith and Nature', *MR*, xv (1954), 288

R. Stephan: 'Hindemith's Marienleben', *MR*, xv (1954), 275

P. Evans: 'Hindemith's Keyboard Music', *MT*, xcvii (1956), 572

279

F. Lion: 'Cardillac I und II', *Akzente*, iv (1957), 126

H. Mersmann: 'Paul Hindemith', *Deutsche Musik des XX. Jahrhunderts* (Bodenkirchen, 1958)

H. Tischler: 'Hindemith's Ludus tonalis and Bach's Well-tempered Clavier', *MR*, xx (1959), 217

V. Landau: 'Paul Hindemith: a Case Study in Theory and Practice', *MR*, xxi (1960), 38

C. Mason: 'European Chamber Music since 1929', *Cobbett's Cyclopaedia* (London, 3/1963), 13

H. Boatwright: 'Paul Hindemith as a Teacher', *MQ*, l (1964), 279

H. Redlich: 'Paul Hindemith: a Reassessment', *MR*, xxv (1964), 241

R. Bobbitt: 'Hindemith's Twelve-tone Scale', *MR*, xxvi (1965), 104

F. Reizenstein: 'Hindemith: some Aspersions Answered', *Composer* (1965), no.25, p.7

W. Thomson: 'Hindemith's Contribution to Music Theory', *JMT*, ix (1965), 54

W. W. Austin: 'Hindemith', *Music in the 20th Century* (London, 1966), 396

T. W. Adorno: 'Ad vocem Hindemith', *Impromptus* (Frankfurt, 1968), 51–87

H. Tischler: 'Remarks on Hindemith's Contrapuntal Technique', *Essays in Musicology: a Birthday Offering for Willi Apel* (Bloomington, Ind., 1968), 175

R. F. French: 'Hindemith's *Mass* 1963: an Introduction', *Words and Music: the Scholar's View ... in Honor of A. Tillman Merritt* (Cambridge, Mass., 1972), 83

E. Padmore: 'Hindemith and Grünewald', *MR*, xxxiii (1972), 190

A. Briner: 'Die erste Textfassung von Paul Hindemiths Oper "Die Harmonie der Welt" ', *Festschrift für einen Verleger: Ludwig Strecker* (Mainz, 1973), 203–41

E. Padmore: 'Hindemith, Weill', *Music in the Modern Age*, ed. F. W. Sternfeld (London, 1973), 100

H.-P. Hesse: 'Paul Hindemith und die Natur der Tonverwandtschaften', *Convivium musicorum: Festschrift Wolfgang Boetticher* (Berlin, 1974), 106

A. U. Rubeli: *Paul Hindemiths a cappella-Werke* (Mainz, 1975)

N. G. Shakhnazarova: *Problemï muzykal'noy estetiki v teoreticheskikh trudakh Stravinskovo, Schenberga, Khindemita* [Problems of musical aesthetics in the theoretical works of Stravinsky, Schoenberg and Hindemith] (Moscow, 1975)

J.-H. Lederer: 'Zu Hindemiths Idee einer Rhythmen- und Formenlehre', *Mf*, xxix (1976), 21

Bibliography

G. Metz: *Melodische Polyphonie in der Zwoelftonordnung: Studien zum Kontrapunkt Paul Hindemiths* (Baden-Baden, 1976)

G. Sannemüller: *Der Plöner Musiktag von Paul Hindemith* (Neumunster, 1976)

A. C. Fehn: *Change and Permanence* [Benn's text for *Das Unaufhörliche*] (Berne, 1977)

D. Rexroth: *Erprobungen und Erfahrungen: zu Paul Hindemiths Schaffen in den Zwanziger Jahren* (Frankfurt, 1978)

I. Bent: 'Analysis, §II, 6', *Grove6*

G. Schubert: 'Kontext und Bedeutung der "Konzertmusiken" Hindemiths', *Hamburger Jb für Musikwissenschaft*, iv (1980), 85

R. Kirkpatrick: 'Recollections of Two Composers: Hindemith and Stravinsky', *Yale Review*, lxxi (1982), 627

Index

283

Index

287